Enchiridion of Practical Microbiology

NIPA® GENX ELECTRONIC RESOURCES & SOLUTIONS P. LTD.
New Delhi-110 034

About the Author

Dr. Anshu Sibbal Chatli is born at Batala, a small industrial town of Punjab. She is a meritorious student through-out her studies. She completed her graduation with a Gold Medal from Guru Nanak Dev University, Amritsar. She completed her Masters and Doctorate in Microbiology from Punjab Agricultural University, Ludhiana with university fellowship. She has worked on biofertilizers which was appreciated by various scientific organizations. She has won many best research paper and poster awards too. She worked as Research Associate in CSIR-Institute of Himalayan Bio-resource Technology, Palampur. She has also worked as a Scientist for three years in **Fast Track Young Scientist Scheme of Department of Science and Technology**, Government of India in Department of Soils, PAU, Ludhiana and taught Under Graduate and Post Graduate courses of Soil Biology and Biochemistry & Soil Microbiology. She worked as Senior **Assistant Professor** in Guru Nanak Girls College, Model Town, Ludhiana for 15 years and taught Biotechnology, Microbiology and Bioinformatics. She was very popular teacher amongst students and awarded with best teacher award. She has guided 10 M.Sc. (Biotechnology) students as a Major Advisor and 4 as Co-Advisor. She has compiled practical laboratory manuals and developed various audio visual power point presentations for the teaching of students. She is a motivator and career counselor to her students. Many of her students are placed in the reputed laboratories and Universities in India and abroad. She was also involved in teaching, assignments and survey projects as a counselor in IGNOU center and elected member of Board of Studies, Punjab University, Chandigarh. She has regularly accompanied the students to various symposia/workshops. She has published 150 papers in various journals of repute. In addition, she has published 9 books and 4 book chapters and 27 articles in the newspapers in vernacular language (Punjabi) on popular science. She has delivered more than 20 invited lectures in various symposia/workshops. She has been awarded with travel grant to attend Conference at Montpelliar, France. She has also worked as Director (Quality Assurance) in Farm Value Foods Private Limited, Ludhiana. She has completed her B.Ed. from Punjabi University, Patiala. She has also worked as Additional Director, DM Group of Institutions, Kararwala, Bhatinda. With her vast experience in research, private and administrative and practical exposure in various laboratories, she has compiled all the practical Microbial techniques in this book. On family front, she is married and blessed with two kids, Aayushi and Samarth.

Enchiridion of Practical Microbiology

Anshu Sibbal Chatli

Director, Farm Value Foods Private Limited
Ludhiana, Punjab
Ex-Additional Director, DM Group of Institutes
Kararwala Rampuraphul, Distt Bathinda
Ex-Senior Assistant Professor, Department of Microbiology
Guru Nanak Girls College, Model Town, Ludhiana
Ex-DST Young Scientist
Punjab Agricultural University, Ludhiana-141004

NIPA® GENX ELECTRONIC RESOURCES & SOLUTIONS P. LTD.
New Delhi-110 034

NIPA® GENX ELECTRONIC RESOURCES & SOLUTIONS P. LTD.

101,103, Vikas Surya Plaza, CU Block
L.S.C.Market, Pitam Pura, New Delhi-110 034
Ph : +91 11 27341616, 27341717, 27341718
E-mail: newindiapublishingagency@gmail.com
Web: www.nipabooks.com

For customer assistance, please contact
Phone: + 91-11-27 34 17 17
Fax: + 91-11- 27 34 16 16
E-Mail: feedbacks@nipabooks.com

ISBN: 978-81-19072-03-3

Composed and Designed by NIPA.

Preface

As we delve deeper into the fascinating and complex world of Microbiology, we are constantly presented with new challenges and opportunities to better understand the microbial world. Microbiological techniques are essential in unlocking the secrets of microorganisms and determining their role in our world.

The development of new and improved Microbiological techniques has opened up a world of possibility for researchers and scientists alike. These methods have provided us with tools to study the activity, expression, and diversity of microorganisms in different environments, from the human body to the depths of the ocean.

In this book, I have compiled a comprehensive guide to the most commonly used Microbiological techniques. I have covered a wide range of topics, from basic laboratory techniques to food Microbiological techniques. Each chapter provides detailed instructions and practical tips based on the authors' extensive experience in the field. The questions are also included to assess the learnings of the students.

I hope that this book serves as a valuable resource for students, researchers and anyone interested in Microbiology. The techniques outlined in this book can be applied to a diverse range of research questions and I believe that they will help advance our understanding of the Microbial world.

I want to thank Almighty for blessings and my husband Dr Manish Kumar Chatli, daughter Aayushi and son Samarth for their support and patience.

Finally, I would like to express my gratitude to Dr Vrinda Sikri who edited the book and provided constructive suggestions to shape the book in present form.

This book will be a handy tool for the graduate and post graduate students of Microbiology, Soil Microbiology, Food Technology, Molecular Biology and quality inspectors of food industry.

May this book inspire new discoveries and a deeper understanding of the microbial world.

Anshu Sibbal Chatli

Contents

1

Basic Laboratory Instructions

1. General Instructions

Instructions to work in the laboratory are given below. Students are advised to read them carefully before they begin their work in the laboratory.

- Apparatus to be used should be neat and clean.
- Graduated pipette should be used for pipetting.
- The small amount of liquid remaining in the pipette should not be blown out while standardizing the pipettes; allowance is always made for this. However, a frosted ring near the top of recently manufactured pipettes indicates that the last drop is to be blown out.

Drainage time of various pipettes

Capacity (ml)	**2.0**	**10.0**	**20.0**	**25.0**	**50.0**
Approximate Time of drainage (seconds)	10	20	28	30	35

- Most measuring devices are made of glass which has a small temperature coefficient, e.g., a soft glass vessel will lead to a change in volume by about 0.003 percent per degree in temperature; with heat resistant glass the change is about one-third of this. As a general rule, the heating of calibrated equipment should be avoided. Rapid cooling can change the glass structure and cause changes in volume.

2. Laboratory Safety

1. Laboratory should be well ventilated and fitted with exhaust fans for effective removal of fumes.
2. Use apron and other devices like gloves, goggles etc. depending upon the material to be handled.
3. Do not add water to acids. Keep acids off skin and protect eyes from spattering. If acids are spilled on the skin, wash off immediately with tap water. Gaseous nitrogen oxides from HNO_3 can cause severe lung damage.

Copious flow of fumes happens when both concentrated HNO_3 and HCI are mixed together.

4. If acid falls on clothes, neutralize the same with few drops of dilute ammonia solution or some other weak alkali solution.
5. If acid spills on the floor or the table, neutralize it with some weak alkali and wipe off with duster.
6. If you happen to suck acid into your mouth during pipetting, wash your mouth quickly with water and then rinse with a weak solution of washing soda.
7. Use fume hood/closet, a local ventilation device; to protect against any type of fumes.
8. Avoid use of equipment's for purposes other than intended.

3. Cleaning Laboratory Wares

The glass and porcelain wares should be thoroughly washed with detergent then these should be extensively rinsed with tap water followed by further rinsing with distilled water.

If a grease film remains after cleaning with detergent, a cleaning solution consisting of sodium or potassium dichromate in concentrated sulphuric acid may be used. After this rinsing is necessary in order to remove the last traces of dichromate ions, which adhere strongly to glass or porcelain surface.

Preparation of cleaning solution: Mix 10 to 15 g of potassium dichromate with about 15 ml of water in a 500 ml conical flask. Add concentrated sulphuric acid slowly and stir to bring the mass into solution.

Note

1. Cleaning solution should be discarded when it acquired green colour of chromium ion.
2. Cleaning solution is most effective when warmed up to about 70ºC. At this temperature it rapidly attacks plant and animal matter and thus it is potentially a dangerous preparation.
3. Spillage, if occurs, should be cleaned with water and wiped off.

4. Laboratory Instructions

1. Remember that you are in a microbiological laboratory and are working with microbes, some of which may be capable of producing disease. Therefore, take extreme care in handling the cultures, slides and other materials, while working in the laboratory.

2. Handover all used slides and cover glasses to the attendant or discard them in the specific container.
3. Avoid spillage of materials or stains.
4. Wash your hands before you come and after you leave the laboratory.
5. In case of any accident, please bring it to the attention of the instructor imemdiately.
6. Take extreme care while you handle the Microscope and its eye pieces.
7. If you have used an oil immersion lens, make sure that you have cleaned the lens with xylene before you return the instrument to the laboratory attendent.
8. Always make it a rule to flame the inoculating needle before and after use. Also, always open a culture tube or flask near the flame.
9. The instruction book is provided so that you can prepare and plan your laboratory work. Therefore, read the instructions carefully before you come to the laboratory. Your laboratory report on the exercise you perform should be submitted for evaluation before performing the succeeding exercise.
10. Use a white laboratory apron and a separate piece of cloth to clean up the workbench.

2

Sterilization Protocols for Media and Glasswares

Introduction

Microbes like all other living beings require suitable nutrients as well as favourable environment like pH, osmotic pressure, temperature, humidity etc. for growth. The substrate/food base which fulfills any of these requirements and supports the growth of an organism is called culture medium. Selection of a medium to grow a particular type of microorganism is extremely important, as the microbes vary in their nutritional requirements. Thus, different media are used for different groups of microbes. The following are the main types of media:

a) Natural/Empirical media: These are made up of natural substrates, such as herbaceous or woody stems, seeds, leaves, corn meal, wheat germ, and oatmeal etc.The exact chemical composition of this media is not known. e.g. Nutrient agar

b) Semi-synthetic media: The chemical composition of this media is partially known. e.g. Potato Dextrose agar/Czapek-Dox agar/Oat meal agar etc,

c) Synthetic media/Chemically defined media: This media is composed from non-natural substances/components of well defined chemical composition. e.g., Davis medium

Procedure

Prepare the following three media namely

1. Nutrient Agar as a general purpose liquid medium for wide variety microorganisms. e.g. Bacteria
2. Kenknight and Munaier's agar for actinomyces species from soil sample.
3. Potato Dextrose Agar/Czapek-Dox Agar for molds & yeasts
4. Composition of these media is as follows:

A. Bacteria

Nutrient agar	
Ingredients	**Quantity**
Peptone	5.0 g
Beef extract	3.0 g
Sodium chloride	5.0 g
Distilled water	1000 ml
Agar	15.0 g
pH	6.8-7.0

B. Actinomyces

Kenknight and Munaier's agar	
Ingredients	**Quantity**
Dextrose	1.0 g
Potassium dihydrogen phosphate (K_2HPO_4)	0.1 g
Sodium nitrate (NaNO$_3$)	0.1 g
Potassium chloride (KCL)	0.1 g
Magnesium sulphate heptahydrate ($MgSO_4.7H_2O$)	0.1 g
Agar	20 g
Distilled water	1000 ml
pH	6.8-7.0

C. Fungi

a) Potato Dextrose Agar	
Ingredients	**Quantity**
Potato (Peeled)	250 g
Dextrose	20.0 g
Agar	20.0 g
Distilled water	1000 ml
Streptomycin	75 mg/l of autoclaved medium

Cut potatoes into small pieces and boil them in one litre of distilled water for 20 minutes. Strain the broth through 3-4 layers of cheesecloth and discard the potato pieces. To this broth add dextrose, agar (with stirring) and water to make volume to 1 litre and maintain the pH 6.8-7.0. Autoclave it.

b) Czapek-Dox agar medium	
Ingredients	**Quantity**
Sodium nitrate	2.0 g
Dipotassium hydrogen orthophosphate	1.0 g
Magnesium sulphate	0.5 g
Potassium chloride	0.5 g
Ferrous sulphate	0.01 g

Sucrose	30 g
Agar	20 g
Distilled water	1000 ml
pH	6.8-7.0
Streptomycin**	75 mg/lt. of autoclaved medium

Functions of ingredients

For the growth of heterotrophic soil bacteria, organic source of carbon is needed which is provided through dextrose, peeled potatoes and beef extracts. Some salts like $NaNO_{3,}$ NaCl, $MgSO_4$ and KCl are added to provide nutrients. The use of K_2HPO_4 and KH_2PO_4 besides providing K and P is that it help in maintaining the pH of the medium by acting as buffer. Peptone and beef extract provide the organic source of N and vitamin B. Streptomycin sulphate in Potato dextrose agar acts as an antibacterial agent. The agar, act as solidifying agent and is the extract of seaweed marine algae viz. *Geledium*, *Gracilarium* etc. Agar will harden a solution into a clear gel similar to ordinary gelatin, but it is different from gelatin in certain respects that makes it much more suitable for use in the laboratory of Microbiology. For example, it is non-nutritive for most organisms, whereas many organisms utilized gelatin (a protein) as a nutrient and as food. Another advantage of agar results from its melting and solidifying temperatures. Incubation of most microorganisms is conducted at the melting temperature of gelatin (25°C). A medium solidified by gelatin will therefore become fluid. Agar liquifies beyond 100°C but once it has molten, it will remain fluid to approximately 40°C, just a little warmer than your body temperature. Both agar and gelatin form a clear transparent material when hard, thus making it possible to see through a reasonably thin layer of this medium.

All the ingredients of these media are weighed and dissolved in small amount of distilled water except agar-agar. Make up the total volume. Adjust the pH of the media with dilute HCl or dilute NaOH. Transfer the media in a pan, now add agar-agar (non-absorbent) and boil with occasional stirring. Pour the media in flasks or tubes and plug them with cotton and sterilize.

Preparation of common microbiological apparatus

The apparatus generally used for microbiological purpose are, pipettes, petri dishes, sterile water blanks and Erlenmeyer flasks etc.

a) **Pipettes:** Pipettes are made up of glass and are graduated with numbers 1, 2, 3; -----10. These are available in different capacities such as 0.1, 0.5, 1.0, 5.0 and10 ml. Pipettes are used for making dilutions, transferring solutions and cultures. These should be sterilized in an oven or autoclave.

Generally, for transferring small volumes, a number of laboratories also use pipettes with disposable tips.

b) **Petri-dishes**: These are covered glass dishes, used to provide flat surface to molten culture medium. Washed, dried and properly assembled petri dishes are wrapped in paper singly or in pairs and sterilized.

c) **Sterile water blanks:** These are used for making dilutions or can also serve as a control treatment in some studies. Dispense the desired amount of distilled water in test tubes or conical flasks. Plug them with cotton and sterilize.

d) **Erlenmeyer Flasks:** These flasks were discovered by Erlenmeyer, hence the name. These are available in different sizes like 100, 250, 500, 1000, 2000 ml. Some of them are graduated and used during preparation of media. A cotton plug is inserted in the mouth of flask and it must be sterilized in autoclave.

Sterilization

Sterilization is done to destroy all living organisms without damaging or altering the substances. Different materials are sterilized differently. The choice of the method depends upon the nature of the material to be sterilized. The following are various methods of sterilization.

a) Dry heat sterilization (hot air, sterilization)
b) Wet heat sterilization (steam sterilization)
c) Chemical sterilization
d) Flame sterilization
e) Filter sterilization
f) Radiation method

a) **Dry heat:** Materials to be sterilized are placed in an oven at 160°C for 2 hours. The dry heat is used principally to sterilize glassware or other heat stable solid materials.

b) **Wet heat**: Sterilization by this method is done by autoclave, in which steam is allowed to form in the inner cylinder by boiling water. The rise in pressure is represented by pressure gauge. The steam pressure raises the temperature up to desired level to kill the microorganisms. The normally required media are sterilized at 15 lbs (pounds per sq inch) pressure at 121°C (250°F) for 15-20 minutes.

c) **Chemical sterilization**: A number of chemicals are used in laboratory to sterilize glassware, hands, chamber etc. Ethyl and Isopropyl alcohols are skin antiseptics. The chemicals such as phenyl-mercuric acid, mercuric-chloride and methyl-bromide are the effective killing agents. Some materials are treated with these chemicals to sterilize them. Nodules are surface sterilized by dipping in 0.1% solution of mercuric-chloride in 70% alcohol (v/v).

d) **Flame sterilization**: Some of the materials are sterilized by making them red hot on the flame. The inoculation needles, forceps etc. are sterilized by this method.

e) **Filter Sterilization**: When certain components of media are thermolabile, the use of heat sterilization is not recommended. Pesticides, hormones, solutions of antibiotics, vitamins, animal serum, tissue extracts, blood plasma etc. are decomposed by heating. Such solutions are sterilized by passing through a membrane filter made up of cellulose- acetate or polycarbonate of a particular pore size ($< 0.045\mu$) to get rid of microbes.

g) **Radiation Method**: This is one of the mildest ways of sterilization, employed for solids, liquids or gases where high temperature is undesirable so it is also called as cold sterilization. Tabletops, chambers etc. can be exposed to Ultra violet light (200 to 320 mμ wave length) for ½ hour. Irradiation with X-rays from Cobalt-60 is used with great advantages. The normally present ultra-violet rays in the sunlight also have a mild sterilization effect.

Reference

Wollum II A. G. (1982) Cultural methods for soil microorganisms. In A L Page, R H Miller and D R. Keeney (ed.). Methods of Soil Analysis, Part II. Chemical and Microbiological properties. Agron. Monograph. 9 (Second Ed): 781-802.

Questions

1. Write short note on various methods of sterilization.
2. Why soil is autoclaved for three consecutive days for its sterilization?
3. Why do you use three different media for growing bacteria, fungi and actinomycetes?
4. Why agar-agar is used as a solidifying agent in the medium?
5. What is the function of Streptomycin sulphate in Potato Dextrose agar media?

3

Isolation and Enumeration of Bacteria By Pour Plate Method

Introduction

Soil is the natural habitat of a number of microbes viz. bacteria, fungi, actinomycetes, algae and many viruses. Out of these, bacteria are morphologically very simple as made up of just one cell of small size, therefore can multiply rapidly. Each viable bacterium at appropriate dilution can develop into a colony. Hence by counting the number of these colonies which develop on a selective growth medium, the number of viable cells of bacteria present in soil can be ascertained. Since the number of the bacteria present in soil is very high, the soil suspension is to be sufficiently diluted to reduce the number of colonies and to prevent their overlapping. The number of colonies formed from a diluted sample multiplied by the extent of dilution (dilution factor) gives the number of viable cells present in the sample examined.

Materials

1. Finely sieved soil
2. Sterile water blanksof 9 ml each
3. Sterile pipettes of 1 ml
4. Sterile petri-dishes
5. Colony counter
6. Compound Microscope
7. Medium: Nutrient Agar (Prepared in Exercise-II)
8. Erlenmeyer flasks

Procedure

1. Separate soil closely adhering to the plant roots after gentle tapping/ shaking and air dry it.
2. Weigh and suspend 1 g of dried soil in 9 ml distilled water blank to obtain 10^{-1} dilution. Transfer 1 ml of suspension to subsequent 9 ml water blank, it establishes a 10^{-2} dilution. From this second dilution, similarly transfer 1 ml quantities to next water blank of 9 ml to provide dilution series ranging from 10^{-3},10^{-4}, 10^{-5} and 10^{-6} dilutions.

3. From the highest dilution prepared (assume 10^{-6}) transfer 1 ml portion of the freshly agitated suspension to each 3 sterile petri-dishes by means of sterile pipette. Make similar transfers from the next lower dilutions (10^{-5} and 10^{-4}) into other dishes. Inoculate 1 ml water from sterile water blank in 3 petri dishes. It will serve as control.
4. Into each seeded petri-dish pour about 20 ml of nutrient agar, which previously has been steamed sufficiently to ensure complete melting and then cooled to 42°C.
5. Rotate the petri-plates clockwise and anti-clockwise to mix soil suspension with medium.
6. Permit the agar to solidify. Then invert them, and place them in an incubator at 28°C for 3-4 days.
7. Upon removal from the incubator, inspect all plates prepared from a single dilution to see whether a proper dilution has been plated and whether a proper dilution effect is apparent. The latter means that a plate from a 10^{-6} dilution should have only approximately 1/10th as many colonies as the plate prepared from 10^{-5} dilution. If there are numerous colonies on the plates, and if a dilution effect is not apparent, contamination has undoubtedly occurred, and all plates should be discarded. The counts in control shall confirm contamination.
8. If the plates from highest dilution show more than 300 colonies, the dilution has been too low, and if those from the lowest dilution show less than 30 colonies the dilutions have been too high. In either case, discard all plates. Select the plates from the dilution at which 30 to 300 colonies have developed.
9. Calculate the results by multiplying average number of colonies per plate for a given dilution by the dilution factor to the volume of sample suspended.
10. The isolated bacteria are purified on Nutrient agar media slants and microscopic observations are made on cell morphology viz. shape, size and arrangement.

Observations and Calculations

No. of bacterial colonies in different dilutions				
Plate (Replication)	**10^{-3}**	**10^{-4}**	**10^{-5}**	**10^{-6}**
1				
2				
3				
Average				

Select the dilution in which the number of colonies falls between 30-300. Reject the other dilutions.

$$\text{Colony forming units per gram of dry soil} = \frac{X \times \text{dilution factor}}{\text{Volume of sample suspended}}$$

X= number of colonies

Precautions

i. Always mark the Petri plates on the base, never on the lid.

ii. The temperature of molten agar while pouring should not exceed 45 ^{0}C.

iii. The care should be taken while mixing the media and sample that it should not spill over the plates.

iv. Sterilize the laminar flow chamber properly before proceeding with the experiment.

v. Always count the plates with colonies between 30-300.

References

1. Gerhardt P., Murray R.G.E, Costilow R.N, Nester E.W, Wood W.A., Kreig N.R. and Phillips G.B. (1981) Manual of Methods for General Bacteriology, Washington: American Society for Microbiology.

2. Murray R.G.E. and Robinow C.F. (1981). Manual of Methods for General Bacteriology, Washington.

3. Starr M. P., Stolp H. Truper H.G., Balows A. and Schlegel H.G. (ed.) (1981). The Prokaryotes. A handbook on habitats, isolation and identification of bacteria. Vol. II. Springer-Verlag New York.

Questions

1. Why Nutrient agar medium is used for viable counts of bacteria?
2. Why media with low energy source is used in this experiment?
3. Which portion of the bacteria in soil is counted by this method?
4. Nutrient agar is of which type media?
5. Name the factors which determine the accuracy of plating method.
6. Why media is cooled to about 42°C for plating?
7. What is the importance of viable counts of bacteria in soil?

4

Isolation and Enumeration of Fungi by Pour Plate Method

Introduction

Fungi constitute an important part of microflora of normal soil. In soils, fungal propagules vary from 2 x 10^4 to 1 x 10^6 per gram soil. The propagule being considered as fungal hyphae, spore, and hyphal fragments which is capable of giving colony on the plate. Most soil forms have a filamentous organisation capable of reproduction by means of spores. Their development is especially favoured by soils having acidic reactions. Soil fungi has significant role in organic matter decomposition, as a plant pathogens and in the interaction with other soil microorganisms. Agar plate count method is commonly used as a measure of fungal population in soil.

Materials

1. Finely sieved soil
2. Sterile petridishes
3. Sterile water blanks 9 ml
4. Sterile pipettes 1 ml
5. Colony counter
6. Microscope
7. Potato Dextrose agar. (prepared in Exercise II)
8. Erlenmeyer Flasks

Procedure

1. Take 1 g of dried soil sample and prepare ten-fold dilution series ranging from 10^{-1} to 10^{-6} as described in Exercise II. For many soils satisfactory dilution range will be 10^{-3} to 10^{-6}.
2. Aseptically pour 1 ml suspension from 10^{-3} to 10^{-6} dilution to each 3 dishes after shaking.

3. Transfer 1 ml aliquot from water blank into 3 petri dishes to serve as control.
4. Add approximately 20 ml of the medium, melted and cooled to approximately 42°C.
5. Mix the soil suspension with the melted agar by rotating or swirling the freshly poured agar.
6. After solidification of agar invert the plates and incubate them at 28°C for 5 to 6 days.
7. Select for counting the dilution range that yielded from 10 to 100 colonies per plate. In the plate, count only those colonies, which are typically filamentous and not bacterial in composition.
8. Convert the average count to a number per gram basis as described in Exercise II. The isolated fungal colonies are purified on Potato dextrose agar slants and are identified.
9. After incubation note the cultural characteristics of the organisms that develop. Record differences in pigmentation, texture, and form of colonies.
10. Examine representative colonies with the low power objective of the microscope, remove a small bit of the colony to a glass slide and examine in a drop of lacto-phenol mounting fluid under a cover slip. Make drawings of the different forms observed.

Observations and Calculations

No. of colonies in each dilution				
Plate No.	10^{-3}	10^{-4}	10^{-5}	10^{-6}
1				
2				
3				
Average				

Select the dilution where the number of colonies are in between 10-100 colonies per plate.

$$\text{Colony forming units per gram in dry soil} = \frac{X \times \text{dilution factor}}{\text{Volume of sample suspended}}$$

X= number of colonies

References

1. Harley J. L. and Waid J. W. (1955) A method for studying active mycelia on living roots and other surface in the soil. Trans Br. Mycol. Soc. 38: 104-18.

2. Kanwar S.S., Gupta M.K. and Punj V. (1997) Laboratory Manual of Basic Microbiology, Department of Microbiology, College of Basic Sciences, Himachal Pradesh Krishi Vishvavidyala, Palampur.

Questions

1. What is the role of fungi in soil?
2. Why Streptomycin sulphate is added to Potato Dextrose agar?
3. What are the drawbacks of plate count method for counting fungi? Which plates should be selected for counting fungi by agar plate method?
4. Name inhibitors, which can allow only the fungus to grow and not other organisms?
5. The fungal colonies developed from spores are counted in Pour Plate method, comment.

5

Isolation and Enumeration of Actinomycetes by Pour Plate Method

Introduction

The actinomycetes are gram positive, non-motile, non-sporulating and non-acid fast bacteria. Sometimes these are called ray fungi and are connecting link between the bacteria and the true fungi. They possess a very fine vegetative mycelium that does not fragment into bacillary or coccoid forms and therefore their colonies on agar are commonly cohesive or tough in consistency. In many soils, the actinomycetes accounts from 10 to 25% of the total count on agar plates. Their number is higher in soils of low moisture content than in wet soils. They are responsible for degrading more resistant carbon material in soil like humus. They are also important flora of manure heaps. Most of these, particularly *Streptomyces*, produce antibiotics and so are of a great value to the pharmaceutical industry.

The counting of actinomycetes population of soil by plate count method is complicated because no cultural substrate or medium have been developed that is highly selective for actinomycetes. The plates commonly contain colonies of fungi and bacteria along with those of actinomycetes. For accurate counts of actinomycetes, the colonies are first differentiated from bacterial and fungal colonies and then they are counted.

Materials

1. Finely sieved soil
2. Sterile petri dishes
3. Sterile water blanks 9 ml
4. Sterile pipettes 1 ml
5. Colony counter
6. Microscope
7. Kenknight and Munaier's agar (Prepared in Exercise II)
8. Erlenmeyer Flasks

Procedure

1. Take 1.0 g of dried soil sample.
2. Prepare serial dilution from 10^{-1} to 10^{-6} (As per Exercise II).
3. Aseptically pour 1.0 ml suspension from selected dilution in 3 plates for each dilution. Maintain control.
4. Pour approximately 20 ml of Kenknight and Munaier's agar to each plate.
5. Gently rotate the petri plates to distribute the soil inoculum throughout the medium. Permit the agar to solidify, invert and incubate the plates at 30°C for 7 days.
6. Select the plates for counting the colonies. Each plate showing at least 30 and not more than 300 colonies are selected. Small radiating colonies of powdery appearance of actinomycetes grow on medium.
7. For the average colony count, calculate the population per g of soil as described.

Observations and Calculations

No. of colonies in each dilution				
Plate No.	**10^{-3}**	**10^{-4}**	**10^{-5}**	**10^{-6}**
1				
2				
3				
Average				

Select dilutions having colonies between 30-300.

$$\text{Colony forming units per Gram of dry soil} = \frac{X \times \text{dilution factor}}{\text{Volume of sample suspended}}$$

X= number of colonies

Reference

Nonomura H., and Ohara Y. (1969). Distribution of actinomycetes in soil. IV. A culture method effective for both preferential isolation and enumeration of *Microbispora* and *Streptosporangium* strains in soi (Part I). *Hakku Kogaku Zasshi* 49: 895-903.

Questions

1. How you will identify the actinomycetal colony?
2. Why fungus and bacteria grow along with actinomycetes?
3. Differentiate between inoculation and incubation.

4. With which number of actinomycetes colonies the plates are selected for recording results?
5. Actinomycetes are the transition between true bacteria and fungi. Comment.

6

Microbial Staining Methods

Introduction

Bacteria are very minute and colourless in their natural environment and are difficult to be seen under compound microscope. To view their exact morphology, various kinds of stains are used. Stains are of three main types.

Acidic stains: These are those stains in which charge on dye ion is negative. e.g. India Ink, Eosin etc.

Basic stains: These are those stains in which charge on dye ion is positive. e.g. Methylene blue, Crystal violet etc,

Neutral stains: These are the salts of dye acid in dye base. e.g. Eosinate of Methylene blue.

Staining Techniques

a) **Simple staining or Monochrome staining**

In this staining technique only one staining solution is used to study morphology and arrangement of bacteria.

Procedure

1. Fix the smear on glass slide by gentle warming over the flame of spirit lamp.
2. Flood the smear with a solution of basic dye such as methylene blue for approximately 1-2 minutes.
3. Gently rinse with distilled water and dry with blotting paper.
4. Examine the smear under oil immersion objective.
5. Bacteria are observed blue or violet depending upon the stain used.

b) **Negative staining or Relief staining**

This method requires the use of an acidic dye such as Nigrosin. The acidic stain will not penetrate the cells due to the negative charge on the surface of bacteria. Thus the unstained cells can be easily observed against the coloured background. This staining method is used to see the shape of bacteria.

Nigrosin stain -- 10.0 g

Water -- 100 ml

Boil for 30 minutes and add 0.5 ml formalin. Filter through filter paper and store 2 ml content in sterile tubes. Place a loopful of the suspension to be examined on a slide. Immediately add an equal amount of the nigrosin solution, mix and spread out in a thin film.

Procedure

1. Place a droplet of Nigrosin on a glass slide.
2. Mix a small sample of bacterial culture into the droplet. Make a thin smear. Let the film dry.
3. Observe the slide under oil immersion.
4. Colourless bacterial cells are observed against a dark field background.

c) Gram staining or Differential staining

Gram staining was developed by a Danish Scholar, Hans Christian Gram in 1884. This technique is employed to visualizethe difference between two types of bacteria: Gram Positive and Gram Negative. The steps required for Gram staining are as below:

1. Applying a Primary stain- Crystal violet

   Solution A- 2.0g Crystal violet in 20 ml 95% alcohol

   Solution B- 0.8 g Ammonium oxalate in 80 ml distilled water.

   Mix solutions A and B
2. Adding a Mordant-Gram's Iodine

   Iodine crystals – 1.0 g

   Potassium iodide – 2.0 g

   Distilled water – 300 ml

   Allow to stand 24 hours for the Iodine to dissolve. It may be necessary to add a few more crystals of Potassium iodide.
3. Rapid decolourization with Ethyl alcohol 95% v/v
4, Counterstaining withSaffranine

   2.5 per cent solution in 95% alcohol ----------- 25 ml

   Distilled water -- 75 ml

The Gram-negative bacteria have high lipid content in cell wall. Thus, a large amount of lipid is easily dissolved by alcohol which makes large sized pores in the cell wall. These large sized pores are also not closed by dehydration effect of alcohol. This results in release of Crystal violet-Iodine complex out of the cell. The cells are counter stained with Saffranine, hence Gram-negative bacteria are pink in colour. In Gram-positive bacteria, the lipid content is very low in cell wall. Small amount of lipid is easily dissolute by Ethanol and hence minute pores are developed in cell wall which can be closed easily with dehydration. Hence, the cells retain Primary stain i.e. Crystal violet stain of purple colour.

Procedure

1. Prepare homogenous bacterial smears from one-day-old broth culture on clean glass slides, air dry and heat fix.
2. Flood the smear with crystal violet stain for1 minute, wash with distilled water, flood with mordant Gram's Iodine solution for 1 minute. Wash the slides with distilled water.
3. Decolourise with 95% alcohol until no violet colour comes off.
4. Counter stain the smear with safranine for 30 seconds.
5. Wash the smear, blot dry and observe under microscope using oil immersion objective.
6. Gram-positive bacteria will appear purple while the Gram-negative pink in colour.

Endospore Staining (Schaeffer-Fulton staining method)

Under the conditions of inadequate nutrition or unfavourable environment for growth, specialized structures called spores are formed. The spores are resistant to heat, drying, freezing, toxic compounds and radiations. They also have high Ca^{++} content and dipicolinic acid (DPA). Most of the members of Bacilli group of bacteria form endospores.

Requirements

1. Starved cultures of *Bacillus cereus*
2. Glass slides
3. Water bath
4. Malachite green (0.5%)
5. Saffranine (0.5%)
6. Blotting paper

7. Oil Immersion
8. Microscope

Procedure

1. Prepare homogenous smears of starved culture of *Bacillus cereus* and heat fix after air drying.
2. Flood the slide with 0.5% malachite green. Put the slide over a water-bath for 5 minutes. Avoid boiling the stain.
3. Wash the slide with distilled water and cover the film with counter stain saffranine.
4. Wash the smear again, blot dry and observe under microscope using oil immersion objective.
5. Spores will appear green while the cytoplasm is stained red.

Capsule Staining (Duguid Staining Method)

A few bacteria viz. *Klebsiella pneumoniae*, *Haemophilus influenzae* etc. form capsules and slime layers. These layers are composed of polysaccharides, a glycoprotein or a polypeptide and is attached to outer cell wall.

Requirements

1. Skimmed milk bacterial cultures of *Klebsiella pneumoniae*, *Haemophilus influenzae*
2. Glass slide
3. India Ink
4. Crystal violet
5. Blotting paper
6. Microscope

Procedure

1 Mix a loopful of bacteria with a small drop of India Ink.

2 Prepare homogenous smear of India Ink mixed with bacteria, air dry and gently heat fix.

3 Cover the smear with Crystal violet for 1 minute.

4 Wash the slide gently with distilled water.

5 Press it with blotting paper and observe under oil immersion objective.

6 The capsules appear as clear zone around the refractive organism with the brownish black background of India ink.

Preparation of Microscopic Mounts and Staining of fungi

The stains required for fungal staining are

1. **Lactophenol:** Mix 20 g Phenol (crystals), 20 ml lactic acid, 40 ml glycerin, in 20 ml water.
2. **Lactophenol-Cotton Blue:** Combine 100 ml lactophenol, 5 ml aqueous solution of cotton blue, and 20 ml glacial acetic acid.

Procedure

- Prepare microscopic mounts in both 50% glycerol and lactophenol-cotton blue by taking a bit of culture growth with the help of sterile needle and forceps.
- Place cover slips gently over the mounts and seal with DPX mountant (a mixture of Distyrene, a plasticizer and Xylene).
- Record microscopic features using high and low power and oil immersion objectives of the fungi.

References

1. Harley J. L. and Waid J. W. (1955) A method for studying active mycelia on living roots and other surface in the soil. *Trans Br. Mycol. Soc.* 38: 104-18.
2. Kanwar S.S.; Gupta M.K. and Punj V. (1997) Laboratory Manual of Basic Microbiology,Department of Microbiology, College of Basic Sciences, Himachal Pradesh Krishi Vishvavidyala, Palampur.

Questions

1. What is the purpose of heating dried bacterial smears before staining with dye?
2. Why is Gram stain of greater value than a simple stain?
3. Why do some organisms stain Gram positive and others Gram negative?
4. What purposes does alcohol serve?
5. What is a mordant?
6. Can we use Methylene blue in place of Nigrosin in Negative staining?
7. Why do bacteria remain unstained in Negative staining?
8. Why Lactophenol-cotton blue is used for staining fungi?

7

Motility Test for Bacteria by Hanging Drop Method

Introduction

The motile bacteria show movements with the help of flagella. The motility of bacteria can be observed by hanging drop method under microscope.

Requirements

1. Bacterial culture
2. Cavity slide
3. Vaseline
4. Coverslip
5. Microscope

Procedure

1. Place a very small drop of 24 hours old bacterial culture in the center of a coverslip.
2. Put a small amount of Vaseline at each cover of the depression side of the cavity.
3. Invert the cavity slide over the coverslip. The coverslip adheres to glass slide. When the coverslip is inverted, the hanging drop is suspended in the well.
4. Observe the preparation under the low and high-power objectives for testing the motility of bacteria.
5. The bacteria that move swiftly across the microscopic field are connoted positive for motility.

Reference

Case L. Christine and Johnson Ted R. (1984) Laboratory Experiments in Microbiology. The Benjamin/Cummings Publishing Company, Inc., Menlo Park, California.

Questions

1. At what age a bacterial culture is actively motile?
2. Why cavity slide is used to observe motility of bacteria?

8

Total Counts of Microorganisms by Haemocytometry

Introduction

A haemocytometer is a single piece of glass with an H-shaped trough forming two counting areas. Hemocytometers are used to determine the total cell count and viability of many different cell types The full grid on a hemocytometer contains nine squares, each of which is 1 mm². Each counting area is defined by a grid system. It has supports that hold the cover glass at the proper distance (0.1 mm) above these areas for the calculation of the concentration per volume of liquid.

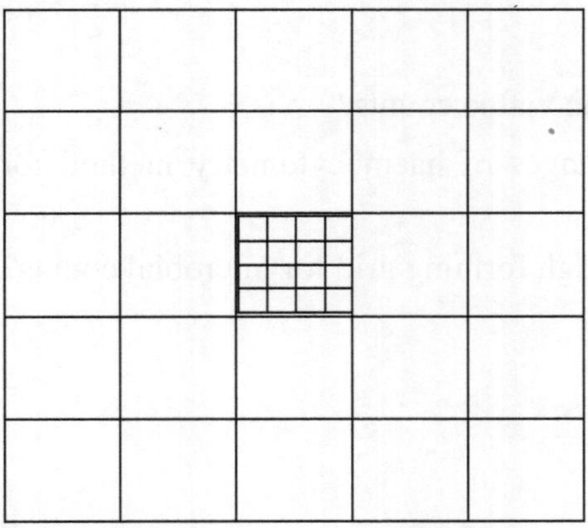

Requirements

1. Inoculum
2. Inoculation needle
3. Test tubes
4. Haemocytometer

Procedure

1. Place a cover glass onto the counting chamber.
2. Place a drop of inoculum at the edge of the cover glass. The drop will be

drawn rapidly into the space between the cover glass and grid.

3. The center square of the grid consists of 25 groups of 16 small squares. Each group is 0.2 mm square. Count the cells in the four corner groups and in the center group. These areas are labelled a, b, c, d and e. To avoid counting a cell twice, count those on a line only when on the top and left line.

Observations and Calculations

Total up the number of cells in all five groups and use in the formula:

Number counted per sub square x 4000 = number/mm^3 x 1000 = number of cells/ml

Counts should be made on both counting areas twice.

Reference

Gupta R.P.; Tewari H.K.; Khanna P.K.; Dhillon G.S. and Kocher G.S. (2004) A Practical Manual for Laboratory Techniques in Microbiology, Department of Microbiology, Punjab Agricultural University, Ludhiana. Pg. 32

Questions

1. Why the direct counts are more than viable counts?
2. List the advantages and disadvantages of haemocytometry method for counting microorganisms.
3. Why haemocytometer is called trough forming grid for microbial counts?

9

Study of Microscope and Use of Stage and Ocular Micrometers

Introduction

A microscope is an optical instrument consisting of a lens or a combination of lenses for making magnified images of minute objects. A simple microscope consists of a single lens of magnified glass held in a frame, usually adjustable and has a stand for holding the object and a mirror for reflecting the light. But a compound microscope consists of two lenses i.e., objective and eye piece respectively. A compound microscope gives much greater magnification than simple microscope. The compound microscope will be extensively used in this course in studying various morphological and staining characters of soil microorganisms.

Terms Frequently Used in Microscopy

Illumination: Microscopes may be illuminated in several ways. Turning the mirror may use daylight, but not direct sunlight, so that light is reflected into the microscope. The concave side is used only when the microscope has no condenser. Simple electric lamps have also been used, placed below the microscope stage instead of mirror. For critical microscopic work, it is necessary to have a lamp with a condensing lens and a diaphragm. The diaphragm is placed in the light path of a lens or objective, to regulate the amount of light that passes through the lens.

Virtual Image: The apparent size and position of the object.

Magnification: The ratio of the apparent size of the object as seen through the microscope to the size of the object as it appears to the unaided eye at a distance of 10 inches. When this ratio applies to the linear dimensions of the object, it called linear magnification and is expressed in units known as diameters, expressed as `X'; e.g., 400X. The compound microscope has two separate lens systems. The one near the specimens, the objective, magnifies the specimen a definite amount. The second, the eyepiece, further magnifies this image (real image), so that image seen by the eye (virtual image) has magnification equal to the product of the magnification of the two lenses. The individual magnification of the objectives for a standard tube length and of the eyepieces is engraved on each such part.

Numerical Aperture (NA): The measurement of the angle of the maximum cone of light that may enter the objective. It is expressed not by the degree of intercepted arc, but by the equation NA = n sin

Where n = refractive index of the material between the object and objective lens

= One half the angular aperture

For a lens with the angular aperture of 111°48', sin = 0.828

Note that oil immersion objective is marked with a NA of 1.25

The greater the NA, the greater the resolving power e.g. an objective with the NA 1.25 has a greater resolving power than an objective with NA, 0.66.

Resolution or Resolving Power (RP): The ability of the microscope objective to separate small details. The resolving power of the lens is its ability to reveal fine details, and is measured in term of the least distance between two point or lines at which they are seen as two rather than as a single blurred object. Most limiting factor to attain good resolution is wavelength of light.

RP = wavelength/2NA

Blue light (0.42-0.40u) permits greater resolving power than red light (0.88-0.88u). Some time UV light (0.20-0.39u) is employed, but this is useful only for photomicrography since the human eye is unable to see objectives illuminated with light of such a small wavelength. When microscope is used at its greatest resolving power, higher magnification will not reveal more detail; such magnification is termed as empty magnification.

Another factor that determines the resolving power is the refractive index of the medium through which light ray's pass. The refractive index of air is 1.0. The refractive index of water and immersion oil is 1.33 and 1.53, respectively. The `n' of glass is 1.52. If light rays after passing through glass pass through air, they will bend before reaching the objective lens, less bending will occur with water than air. Hence maximum angular aperture (the angle of greatest divergence of light rays which objective lens can collect) of the lens will not be realised, which result in poor illumination and decreased resolving power. Consequently with oil immersion objective, space between object and lenses is filled by a medium having 'n' essentially as that of glass. The highest power objective lens will not yield clear image except obtained by use of a condenser and an objective with same NA. If NA = 1.25 and blue light of wavelength of W in use, we can calculate very easily the resolving power of the objective as

RP = 0.48/2x1.25 = 0.19μ

This means that under most ideal conditions this oil immersion objective is capable of distinguishing two object separated by a distance of only 0.11 μ. The most expensive lens does not resolve better than 0.15u with visible light. With UV light it is possible to attain a `PP' of less than 0.11 μ.

Working Distance: The distance between the front of the objective, when it is focused on a thin preparation and top of the glass cover. Most objectives are corrected for a cover glass thickness of 0.18 mm. To prevent water from touching the objectives and to provide a flat surface so that the light can enter the objective, cover glass should always be used. If no cover glass is used with 40x, high dry objective, the image will be hazy.

Depth of Focus: The thickness of the specimen, which may be seen in focus at one time. The greater the magnification and the NA; the thinner will be the layer in focus at one time. The longer focal length of lenses of less magnification will generally be more satisfactory for the study of the general arrangement of the specimen.

Parts of Microscope

1. Mirror: The mirror turns in two planes to allow light rays to be directed along the optical axis of the instrument. It has two sides concave and plane. The concave side should never be used with sub stage condenser, since the latter is designed for use with parallel rays of light, which are reflected from the plane mirror.
2. The sub stage Abbe condenser and Component Iris Diaphragm: are designed to regulate the amount of light passing into the optical system of microscope. The condenser provides uniform illumination by focusing the light as a cone of oblique rays covering on the same plane as the objective to be examined. It is mounted on a rack and pinion to permit vertical movement. Precise focusing of light is essentially important for adequate illumination when the oil immersion objective is used. It is necessary to adjust the intensity of light illumination by opening and closing the iris diaphragm. Generally, it is best to leave the condenser in raised position when using the oil immersion.
3. Inclination Joint: is located between the base and body of microscope. Do not tilt the microscope for examination of wet preparations.
4. The objective Lens: are mounted on a revolving nosepiece. Each lens is marked with NA in small figures and with equivalent focus and magnification in larger figures. The low power lens is usually marked as '16 mm 10x' ; to indicate an equivalent focal distance of 16 millimetres

and a magnification of 10 times. Similarly high dry lens is marked as 4 mm, 40x' and oil immersion objective lens as 1.9 mm, 100 x. A drop of oil is placed on the area of glass slide, which contains the object to be examined under oil immersion objective. The lens is placed in direct contact with the oil. After use the lens should be wiped off with lens paper not with your handkerchief or any other cloth or tissue. When oil is used the lens should be wiped off with lens paper dampened with xylene or chloroform and lens is dried with clear piece of lens paper.

5. Ocular lens or Eyepiece: magnifies the real image formed by the objective lens. Generally, a microscope is equipped with a 10x-eye piece. This ocular is best for general purposes. However, 5x and 15x are also available.
6. The coarse and fine adjustment wheels: are used for raising and lowering body tube. Always focus upward with coarse adjustment wheel.

Materials

1. Microscope, immersion oil, lens paper, glass slides, cover slips
2. Cotton fibre
3. Stained slides of microorganisms
4. Stage and ocular micrometer

Procedure

A. Operation of microscope

1. Grasp the microscope by the arm firmly with one hand and steady it at the base with the other hand. Always carry it in this manner to form its base. It must be kept in upright position. Set it down carefully on the table.
2. Bring the low power objective in position. A distinct click will be heard and felt when the objective lens is in its position.
3. Raise the stage condenser all the up and close the iris diaphragm at least half way. Lower the body tube with coarse adjustment wheel until the end of the low power objective is approximately cm from the upper lens of the condenser.
4. Looking through the eyepiece turn the plane mirror towards the source of light and manipulate the mirror to a position where maximal illumination makes it apparent that the light rays have been centered in the optical axis. Lower the sub stage condenser until the intensity of illumination is comfortable for observation.
5. Position slide by use of clip on the mechanical stage. The area to be examined should be immediately over the condenser lens. Focus upward

with the coarse adjustment wheel until the object in approximate focus, then switch to fine adjustment wheel to bring the object in to sharp focus. Rotate the eyepiece. How much of the material moves with eyepiece? This is the best way to determine whether or not the ocular lens is dirty. Clean with lens paper if necessary.

6. When suitable field for detailed observations has been found, raise the condenser and body tubes and turn the high dry objective in position. If your microscope is par-focal, the position of the body tube need not be altered when switching objective lenses.
7. Lower the body tube until the high dry lens almost touches the slide or cover slip. Rack upward slowly with coarse adjustment wheel until a rough focus is obtained and focus sharply with fine adjustment wheel. Check your illumination and adjust the diaphragm for greater or lesser illumination.
8. Raise the body tube and add a drop of immersion oil to the slide directly below the objective lens. Slowly lower the body tube until the oil immersion lens makes the contact with the oil. Focus upward with the fine adjustment until the object is in good focus. The illumination may be adjusted by opening or closing the iris diaphragm. The best illumination is not bright. Keep the condenser at its upper stop when using the oil immersion objective.
9. After completing the microscopic examination, clean the oil immersion lens with lens paper or if necessary, with xylene dampened lens paper. Set the low power objective in position and lower the body tube all the way to the stop with coarse adjustment wheel. Do not run the tube down. With added precautions rack the condenser down. The other objective lenses and the condenser should never be cleaned with xylene. Always use dry lens paper on these lenses. The sub-stage condenser top lens if dirty may be cleaned with xylene dampened lens paper and subsequently wiped with dry lens paper.

B. To measure the area of microscopic field at different magnifications

1. Stage micrometer is a device to measure the size of microscopic objects. It is a foot rule in which one mm is divided into 100 divisions. Each division represents 10 μm.
2. Place the stage micrometer on the mechanical stage.
3. Focus 10xobjective over the scale and count the number of divisions visible from end to end of the field.

4 Focus at 40x objective over the scale and count the number of divisions visible from end to end of the field.

5 Put a drop of immersion oil on the stage micrometer and focus oil immersion lens and count the number of division visible from end to end.

6 Calculate the area of microscopic field at 10xX, 40X and 100x10sX magnification.

C. Calibration of ocular micrometer

1. Ocular micrometer is a device used to measure the size of the microscope objects. It has unknown graduations marked on it, thus it is standardized by comparing with the stage micrometers.
2. Put the ocular micrometer disc in the ocular lens from upside.
3. Focus at 10X objective.
4. Match the division of stage and ocular micrometers and calculate the length of each division of ocular micrometer in μm.
5. Similarly focus the 40X, and 100X objectives and match the division of stage and ocular micrometers. Calculate the length of each division of ocular micrometer in μm.

Observations and Calculations

A. **Area of microscopic fields at different magnifications**

10x10X magnification

Number of stage micrometer divisions covering the diameter of microscopic field (MF) = A

Diameter (D) of MF = Ax10 μm

Radius of MF = Ax10/2 = 5A μm

Area of MF = $r^2 = \pi (5A)^2$ μm^2

40x10X magnification

Number of stage micrometer divisions covering the diameter of MR = B

Diameter of MF = Bx10 m

R of MF = Bx10/2 = 5B m

Area of MF = $(5B)^2$ μm^2

100x 10X magnification

Number of stage micrometer divisions covering diameter of MF = C

D of MF = Cx10 m
R of MF = Cx10/2 = 5C μm
Area of MF = $(5C)^2$ μm^2

B. Calibration of ocular micrometer

10x10X magnification

Number of ocular (OM) divisions matching with 50 divisions of Stage micrometer (SM) = X
Length of X division of OM = 50 x 10 μm
Length of one division of OM = 50 x 10/X μm

40x 10X magnification

Number of ocular divisions matching with the 10 divisions of SM = Y
Length of Y divisions of OM = 10x10 μm
Length of one division of OM = 10x10/Y μm

100 x 10 magnification

Number of ocular divisions matching with 10 divisions of SM = Z
Length of Z divisions of OM = 10x10 μm
Length of one division of OM = 10x10/Z μm

References

1. Benson Harold, J. (1990) Microbiological Applications, A Laboratory Manual in General Microbiology, Part I Microscopy. Pg. 22-24.
2. Case L. Christine and Johnson Ted R. (1984) Laboratory Experiments in Microbiology. The Benjamin/Cummings Publishing Company, Inc., Menlo Park, California. Pg. 15-16.
3. Doetsch R.N. (1981) Determinative Methods of Light Microscopy. In: Manual of Methods for General Bacteriology, Washington.
4. Spencer M. (1982) Fundamentals of Light Microscopy. New York: Cambridge University Press.

Questions

1 Differentiate between the resolving power and magnifying power.

2 What magnification would result by use of a 15X ocular with three objectives found on your microscope?

3 Why is 16mm objective placed in position while carrying or storing the microscope?

4 Why is oil necessary while using the 1.8-mm objective?

5 What are the units of measurement used in microscopy? It is what part of a) centimeter b) millimeter c) angstrom units?

6 Which objective requires more light for effective use and why?

7 Is there a lower limit to the size of objectives we can see with a microscope? Explain.

10

Isolation and Culturing of Photosynthetic Bacteria

Introduction

Photosynthetic bacteria are autotrophs and anaerobes. These bacteria produce Hydrogen sulphide. These are predominant in the bottom of stagnant water enriched with mud.

Requirements

1. Mud sample
2. Enrichment medium of Photosynthetic bacteria or salt solution (1.0g/lt. of $(NH_4)_2SO_{4,}$ 1 g; $K_2HPO_{4,}$ 1 g; $MgSO_4$,0.2g; NaCl, 5g; pH 7.2)
3. Sterilized bottle with plug
4. Fluorescent tube

Procedure

1. Put 10 ml of mud sample in 250 ml sterilized bottle and plug it.
2. Add 90 ml of Enrichment media (salt solution) in it.
3. Shake the bottle vigoursly to mix all the contents and replug it.
4. Put the bottle approximately 12 cm away from fluorescent tube for 20 days at 30°C temperature.

Results and Observations

If the medium turns red, it will indicate purple bacteria and if green then green photosynthetic bacteria.

Reference

Murray R.G.E. and Robinow C.F. (1981). Manual of Methods for General Bacteriology, Washington.

Questions

1. Why stagnant muddy water is required for the growth of photosynthetic bacteria?
2. Why do different photosynthetic bacteria show different colours in water sample?

11

Isolation of *Rhizobium* from Root Nodules

Introduction

Rhizobia are Gram-negative soil bacteria , aerobic nitrogen fixers and adhere to the roots of leguminous plants. Fixed nitrogen is assimilated to proteins and is transported along the plants. A part of proteins is also released in rhizosphere region.

Requirements

1. Yeast Extract Mannitol Agar (YEMA)
2. Sterile saline
3. $HgCl_2$ solution (1%)
4. Ethanol (70%)
5. Root nodules of Mung bean
6. Water blanks
7. Pipettes
8. Petri plates
9. BOD Incubator
10. Burner

Yeast Extract Mannitol Agar

Ingredients	Quantity
K_2HPO_4	0.5g
$MgSO_4.7H_2O$	0.2g
Sodium chloride	0.1g
Mannitol	10g
Yeast Extract	0.4g
Distilled water	1000ml
Agar	18g
pH	6.8-7.0

Procedure

1. Procure healthy nodules of Mung bean and wash them in sterile solution followed by immersion in 1% $HgCl_2$ solution for 5 minutes.
2. Transfer nodules in saline solution and treat with Ethanol (70%) for 2-3 minutes.
3. Wash the nodules 5-6 times with distilled water and blot drt them
4. Crush the nodules (3-4) with sterile glass rod and prepare serial dilutions upto 10^{-6}. Further, plate $10^{-\backslash 4}$ - 10^{-6} dilutions on YEMA plate.
5. Incubate the petri plates at 28°C for 48 hours and observe the colonies.
6. Observe the bacterial colonies.

Results and Observations

Rhizobial colonies are white or translucent and gummy.

References

1. Sharma K. (2008). Manual of Microbiology- Tools and techniques. Second Edition, Anne Books India, New Delhi (India)
2. Benson H.J. (2002). Microbiological Applications; a Laboratory Manual in General Microbiology, McGraw Hill, Eighth Edition, USA.

Questions

1. What is the role of $HgCl_2$ and Ethanol during isolation of *Rhizobium*?
2. What is the difference between Bacteroides and Rhizobial cells?

12

Enumeration of Rhizobium by MPN Method

Introduction

As Rhizobia are slow growers so their enumeration by plating techniques is less recommended. Rhizobia can be enumerated in the presence of a large number of other organisms by most probable number (MPN) method. Soil containing Rhizobia is diluted in liquid and dilutions are used to inoculate roots of legumes. The presence of Rhizobia in a given experimental unit, inoculated by a given dilution, is indicated by the formation of at least one nodule on a root. Thus, the method involves the growth of legumes under conditions such that a single rhizobia cell can multiply and forms at least one nodule on the root.

Most Probable Number Method: This method permits estimation of population without an actual count of single cells or colonies. The method is based on the probability of the growth of the microbe concerned, thus the number is calculated on the basis of probability. It is also called dilution method.

MPN technique is based on the determination of the presence or absence of microorganisms in several individual aliquots of many consecutive dilutions of the soil or other material. The prerequisites or assumptions for using this method for counting any organism are:

a) Micro-organisms in question must be able to bring about some characteristic growth or recognizable transformation in the medium in which it is inoculated,

b) There should be no antagonism

c) The positive growth means that at least one living microbial cell was present initially in the inoculum.

On the basis of probability theory, it is possible to calculate the number of microorganisms from the number of positive and negative tubes; those have received certain quantity of inoculums from different dilutions. The general equation of Halvorson and Ziegler (1933) can be used for calculating the MPN of micro -organisms.

$$a_1n_1 + a_2n_2 + a_2n_3 = \frac{a_1p_1}{1-e^{-a1x}} + \frac{a_2p_2}{1-e^{-a2x}} + \frac{a_3p_3}{1-e^{-a3x}}$$

Where the subscripts 1, 2, 3 refer to three consecutive dilutions, a_1 a_2 and a_3 refer to relative quantity of the original sample in the inoculums applied: n_1, n_2 and n_3 are the number of tubes inoculated P_1, P_2 and P_3 are the number of tubes showing growth; e is the base of natural logarithm (2.718); and x is the most probable number of organisms present in the quantity of inoculums added in second dilution.

Multiplying x by the dilution factor gives the MPN of the organisms in the original sample. Solution of the equation for x, each time is time consuming. Thus keeping all other variables constant, the equation has been solved for x, for different value of P_1, P_2 and P_3. These values have been given in table 1. Now P1 is the number of positive tubes in the least concentrated dilution in which all the tubes are positive or the greater number of tubes is positive. P_2 and P_3 present number of positive tubes in next two higher dilutions. Then find the row numbers in the table in which P_1 and P_2 corresponds. Follow the rows headed by observed value of P_3. Note the value of X where the two rows intersect. Multiply this number by appropriate dilution factor and obtain the MPN in original sample.

Sample: Consider the instance in which a 10-fold dilution with 5 tubes at each dilution yielded the following numbers of positive tubes after incubation.

5 at 10^{-5}, 5 at 10^{-6}, 3 at 10^{-7}, 1 at 10^{-8} and 0 at 10^{-9} in the series $P_1 = 5$, $P_2 = 3$ and $P_3 = 1$

For this combination of P1, P2 and P3 value of X for table is 1.1 or 1.1 organisms are present in the inoculums in 10^{-7} dilution.

Multiply this number by the dilution factor (10^7) gives 11 million as the MPN of organism in the original sample.

Medium

Nutrient solutions for growing plants with and without N+

Stock solution	Chemical	Concentration
Without N		
1.	KH_2PO_4	34 g/litre
2.	$MgSO_4$	123 g/litre
3.	K_2SO_4	65 g/litre
4.	$CaSO_4.2H_2O$	0.1 g
5.	$FeCl_3.6H_2O$	1.4 g/litre
	Na_2H_2EDTA	1.7 g/litre

6.	KCl H_3BO_3 $MnSO_4.H_2O$ $ZnSO_4.7H_2O$ $CuSO_4.5H_2O$ H_2MoO_4 (88% MoO_3)	0.75 g/litre 124 mg/litre 67 mg/litre 46 mg/litre 10 mg/litre 2 mg/litre
With N		
3A●	KNO_3	101 g/litre
	$(NH_4)_2SO_4$	133 g/litre

+ The nutrient solution is a modification of that of Munns (1968). Adjust to pH 6.5 by adding 1N KOH.

Dissolve the two chemicals separately in 500 ml of water before combining. For complete nutrient solution, substitute solution 3A for 3.

Materials

1. 200 ml long glass tubes
2. Racks/ tube holders
3. 100 ml Erlenmeyer flasks
4. Test tubes of 15 ml capacity
5. High quality seeds of legume
6. Nitrogen free nutrient solution
7. Drinking straws
8. Ethanol solution: 70% saturated solution of Calcium hypochlorite
9. Rhizobia culture
10. Rhizobia populations to be enumerated

Procedure

1. Determine the moisture content of fresh soil.
2. Prepare serial dilutions up to 10^{-7}
3. Inoculate a batch of 5 tubes containing 9 ml medium of dilution 10^{-1}, in each tube. Similarly, a batch of 5 tubes of the same medium is inoculated with higher dilutions up to 10^{-7}. Inoculate 5 tubes (control) with water blank.
4. Incubate the tubes for 6 days at 28°C.
5. Place the Rhizobia free 200 ml glass tubes along with autoclaved 50 ml nutrient agar media in a rack/ holder.

6. Rhizobia are killed on the legume seed surface by soaking them in a bactericidal solution for 2 minutes. Rinse the Rhizobia free seed with sufficient water to remove any residual bactericide. Germinate the seeds on filter paper by providing moisture conditions with sterilised distilled water.
7. Put the germinated seed in 200 ml glass tubes containing sterilised medium.
8. Put the plastic drinking straws in these tubes so that they extend half way down the tube 1 to 2 cm above it. The straws facilitate watering and prevent contamination of tubes.
9. Include both positive and negative controls. The positive control consists of plants inoculated with a culture of Rhizobia while negative controls are the tubes that are uninoculated.
10. When seedlings are established, inoculate by adding 1ml of appropriate serial dilution (inoculated in 9 ml broth) onto the root system.
11. Proper watering is required for the establishment of plantlets.
12. Grow the plants approximately for 1 week after nodules are observed on the positive controls. Generally, 3 weeks after inoculation is sufficient for large seeded legumes. Record each tube as positive (one or more nodules present) or negative (no nodules present).
13. Table for tenfold dilution series has been presented inoculating five replications per dilution.

Observations and Calculations

Dilutions	Tubes inoculated	Positive_tubes	Negative_tubes
Control	5		
10^{-1}	5		
10^{-2}	5		
10^{-3}	5		
10^{-4}	5		
10^{-5}	5		

Select P_1, P_2 and P_3

	Value	Dilution
P_1		
P_2		
P_3		

Value from table = x

Rhizobia population per gram of wet soil = X x dilution factor of P_2

Rhizobia population per gram of dry soil =

Table 1 : Most probable number for use with 10-fold dilutions and 5 tubes dilution (Cochran, 1950)

P_1	P_2	Most probable number for indicated value of P_3					
		0	**1**	**2**	**3**	**4**	**5**
0	0		0.018	0.036	0.054	0.072	0.09
0	1	0.018	0.036	0.055	0.073	0.091	0.09
0	2	0.037	0.055	0.074	0.092	0.110	0.13
0	3	0.056	0.074	0.093	0.110	0.130	0.15
0	4	0.075	0.094	0.110	0.130	0.150	0.17
0	5	0.094	0.110	0.130	0.150	0.170	0.19
1	0	0.020	0. 40	0.060	0.080	0.100	0.12
1	1	0.040	0.061	0.081	0.100	0.120	0.14
1	2	0.061	0.082	0.100	0.120	0.150	0.17
1	3	0.083	0.100	0.130	0.150	0.170	0.19
1	4	0.110	0.130	0.150	0.170	0.190	0.22
1	5	0.130	0.150	0.170	0.190	0.220	0.24
2	0	0.045	0.068	0.091	0.120	0.140	0.16
2	1	0.063	0.092	0.120	0.140	0.170	0.19
2	2	0.093	0.120	0.140	0.170	0.190	0.22
2	3	0.120	0.140	0.170	0.200	0.220	0.25
2	4	0.150	0.170	0.200	0.230	0.250	0.28
2	5	0.170	0.200	0.230	0.260	0.290	0.32
3	0	0.078	0.110	0.130	0.160	0.200	0.23
3	1	0.110	0.140	0.170	0.200	0.230	0.27
3	2	0.140	0.170	0.200	0.240	0.270	0.31
3	3	0.170	0.210	0.240	0.280	0.310	0.35
3	4	0.210	0.240	0.280	0.320	0.360	0.40
3	5	0.250	0.290	0.320	0.370	0.410	0.45
4	0	0.130	0.170	0.210	0.250	0.300	0.36
4	1	0.170	0.210	0.260	0.310	0.360	0.42
4	2	0.220	0.260	0.320	0.380	0.440	0.50
4	3	0.270	0.330	0.390	0.450	0.520	0.59
4	4	0.340	0.400	0.470	0.540	0.620	0.69
4	5	0.410	0.480	0.560	0.640	0.720	0.81
5	0	0.230	0.310	0.430	0.580	0.760	0.95
5	1	0.330	0.460	0.640	0.840	1.100	1.30
5	2	0.490	0.700	0.950	1.200	1.500	1.80
5	3	0.790	1.100	1.400	1.800	2.100	2.50
5	4	1.300	1.700	2.200	2.800	3.500	4.30
5	5	2.400	3.500	5.400	9.200	16.000	

13

Isolation of Nitrifiers on Silica Gel

Introduction

Biological nitrification is a two-phase process, which consists of the oxidation of ammonia-nitrogen to nitrite-nitrogen by species of genera *Nitrosomonas*, *Nitrosospira* and *Nitrosogloea*, and the subsequent oxidation of nitrite-nitrogen to nitrate-nitrogen by species of the genera *Nitrobacter* and *Nitrocystis*. The following equations explain the reaction.

(*Nitrosomonas* group)

$2NH_3 + 3O_2$™ ⟶ $2\ HNO_2 + 2\ H_2O$ + 79 calories of energy

(*Nitrobacter* group)

$HNO_2 + \frac{1}{2}\ O_2$™ ⟶ HNO_3 + 21.6 calories of energy

Kluyer and Donker accounted for the following reaction sequence.

NH_3 ⟶	NH_2OH ⟶	$H_2N_2O_2$ ⟶	HNO_2 ⟶	HNO_3
(Ammonia)	(Hydroxylamine)	(Hyponitrousacid)	(Nitrous acid)	(Nitric acid)

The importance of nitrifying organisms in soil is based on their capacity to produce nitrate, the major source of nitrogen assimilated by higher plants. The rate of nitrification is dependent on such factors as concentration of NH_4^+, temperature, moisture, aeration and pH. Bacteria have their optimal activity at an alkaline reaction. Below pH 7.0 the process declines. However, the process is rapid under suitable conditions.

Silica gel is one of the most useful culture solid media. The common difficulty experienced in its successful preparation is the lack of precision. Chemically, silica gel is a colloid, distribution of a solid in a liquid. The latter is completely trapped in the interstices of the former. The following equation explains an inorganic gel when an acid is added to a concentrated solution of sodium silicate (water-glass):

$$(2Na^+ + SiO_3^{\ 2-}) + HOH + 2\ (H^+ + Cl^-) = H_4SiO_4 + 2\ (Na^+ + Cl^-)$$

The molecules of the highly insoluble silicic acid undergo condensation with the splitting out of water between adjacent molecules, such as:

```
        H            H
        O            O
HO -Si – OH –H-O-Si - OH
        O            O
        H            H
```

Materials

1. Cultures of Experiment
2. Bottles-2, 35 ml, of the medium for *Nitrosomonas*
3. Bottles, 35 ml, of the medium for *Nitrobacter*
4. Sterile petri dishes
5. Sterile pipettes
6. Plates of silica gel

Procedure

Preparaton of Silica Gel Plates

1. Liquid media recommended for the isolation of (1) *Nitrosomonas* and (2) *Nitrobacter* spp.
2. Hydrochloric acid, 30% v/v
3. Solution of sodium silicate:
   $NaSiO_3 . 9H_2O$ ……. 20.0% (original pH 10)
   Or
   $NaSiO_3$ anhydrous ….8.5%
4. Acid solution:
   (1) or a (2), as above

10 ml *Nitrosomonas*	10 ml *Nitrobacter*
150 ml Sodium silicate	150 ml Sodium silicate
Bromo thymol blue (0.04%) 2.5 ml	Bromo thymol blue (0.04%) 2.5 ml
Sterile Calcium carbonate ½ g	Sterile Calcium carbonate ½ g

Pipette 15 ml of this solution into a beaker and adjust pH to 7.4 with 30% HCl. Immediately pour the content of the beaker into a petri dish. The gel should set in about 2 minutes. The plates should be dialyzed against liquid medium kept stirred by a stream of bubbles. The plates may be sterilized at 10lbs. Pressure for 15 minutes or the surface may be flamed.

Enrichment Culture

1. Take 100 g of soil in two beakers of 500 ml capacity.
2. Add 10 ml each of *Nitrosomonas* and *Nitrobacter* medium in both these beakers and cover them with Aluminium foil.
3. Incubate these soil samples at 28-30°C for 2 weeks.

Isolation of the Nitrifiers

1. Cover the surface of 2 silica gel plates with 0.5 ml of the ammonium sulphate solution and 2 Plates with the nitrite medium.

   Incubate the plates in a 37°C incubator, or a warm oven for 1-2 hours to permit evaporation of the excess liquid.
2. Serially dilute the incubated enrichment culture from 10^{-1} to 10^{-6} dilution.
3. Spread 0.1 ml of suspension from 10^{-1} to 10^{-6} each in triplicates on silica gel plates and spread it with the help of spreader.
4. Incubate the plate at 25-30°C for 1-2 weeks. Examine the plates under low and high power.
5. The colonies of *Nitrosomonas* spp. are very small, compact, more or less sharply defined and tend to be brownish. The colonies of *Nitrobacter* spp. are tiny and dense.

Reference

Allen O.N. and Allen A. (1957) Experiments in soil bacteriology. 3rd ed. Burgess Publishing Co., Minneapolis, USA.

Questions

1. Characterize the nitrifying bacteria and write the equation for nitrification.
2. What is heterotrophic nitrification? Which organisms are involved and is it likely that they contributed to the results obtained in this exercise?
3. Is nitrification a beneficial process, Explain?
4. Why silica gel is used for the isolation of nitrifiers?

14

Enumeration of Phosphate Solubilising Microorganisms in Soil

Introduction

Phosphate solubilising microorganisms include different groups of micro-organisms such as bacteria, fungi and actinomycetes, which convert insoluble phosphatic compounds into soluble form. The species of *Pseudomonas*, *Micrococcus*, *Bacillus*, *Flavobacterium*, *Penicillium*, *Fusarium*, *Sclerotium*, *Aspergillus* and others have been reported to be active in bioconversion. Such bacteria and fungi can grow in media where $Ca_3(PO_4)_2$, $FePO_4$, $AlPO_4$, apatite, bone meal, rock phosphate or similar insoluble phosphate compounds are the sole source of phosphate. Such microorganisms not only assimilate phosphorus but they also cause a large portion of soluble phosphate to be released in quantities in excess of their own requirements. The solubilisation is not restricted to calcium salts but iron, aluminium and other phosphates are also acted upon. The solubilisation of phosphate in soil is generally due to acidification, chelation and exchange reactions.

The most efficient bacterial isolates, which were identified, are *Pseudomonas striata*, *P. rathonis* and *Bacillus polymyxa*.

The population of PSM is more in rhizoplane followed by rhizosphere and non-rhizosphere as in rhizoplane and rhizosphere the plant roots produce phenolic substances.

Materials

1. Sterlised Petri Plate
2. Sterlised Water Blanks of 9 ml each
3. Soil
4. Pikovskaya's agar medium

Reagent	**g/lt.**
Glucose	10.0
$Ca_3(PO_4)_2$	5.0

$(NH_4)_2 SO_4$	0.5
NaCl	0.2
$MgSO_4.7H_2O$	0.1
KCl	0.2
Yeast extract	0.5
$MnSO_4$	Trace
$FeSO_4.7H_2O$	Trace
Agar	15.0
Distilled water	1000 ml
pH	6.8-7.0

Preparation of tricalcium phosphate suspension

Stock suspension of 2.5 per cent $Ca_3(PO)_2$ is prepared in 0.5 per cent gum Arabic solution and pH adjusted to 8.4 and is sterilized. For preparation of plates, 10 ml of stock suspension of $Ca_3 (PO)_4$ is added especially to the 90 ml sterilized medium.

The medium is supplemented with 40 mg/litre Cycloheximide for development of bacterial and 75 mg/litre streptomycin sulphate for development of fungal colonies.

The PSB can be isolated by serial dilution pour plating.

Procedure

1. Take 1.0 g of air-dried soil sample in a sterile test tube containing 9 ml of sterile distilled water.
2. Shake it for 20 minutes on a mechanical shaker with a magnetic stirrer.
3. Withdraw 1 ml of the above stirred mixture and serially dilute it to 10^{-2}, 10^{-3}, 10^{-4}, 10^{-5}, 10^{-6} dilutions by pouring in 9 ml sterile water blanks.
4. Pour 1.0 ml of required dilution in a sterile petri plate and add molten media cooled to 42^{0}C.
5. Gently swirl the petri plate to mix the media and soil dilution.
6. The colonies forming clear halos on Pikovskaya's agar were considered to be those of the phosphate solubilisers. The colony counts for phosphate solubilisers were made following an incubation of 3 days at 28^0 C.

Observations

1. Cell size
2. Cell shape
3. Gram Reaction in bacteria/ Lacto-phenol cotton blue stain in fungi
4. Cell movement in bacteria

Colony counts on each petridish					
Replications	**Control**	**10^{-1}**	**10^{-2}**	**10^{-3}**	**10^{-4}**
1					
2					
3					
Total					
Mean					

Number of cells per *g* dry soil = Mean number of colonies - Mean No. of colonies in control x dilution factor = Y

Reference

Kannan N. (2002) Laboratory Manual in General Microbiology. Panima Publishing Corporation.

Questions

1. How P- solubilising microbes can solubilise Phosphate?
2. Describe the mechanism of P solubility by microorganisms when tricalcium phosphate is added to the nutrient medium.
3. In acidic soils which sources of inorganic phosphate are being solubilised?
4. What are the sources of phosphate solubilised under alkaline soils?
5. Why the population of PSM is more in rhizoplane and rhizosphere than that of non-rhizosphere?
6. What are the advantages of Phosphate solubilisers over chemical phosphorus fertilizers?

15

Identification and Enumeration of Mycorrhizal Infection in Roots and Soil

Introduction

Mycorrhizal fungi are ubiquitous in soils throughout the world and form symbiotic relationships with the roots of most terrestrial plants. The infection of roots of most flowering plants by symbiotic fungi and the transformation of such roots into unique morphological structures are called Mycorrhiza. Over the years it has been convincingly demonstrated that inoculation of agricultural crop plants and forest trees with mycorrhizal fungi can significantly stimulate their growth and development. Such probiotic effects are more pronounced in nutritionally poor soils which occur in very large areas of the arid and semi-arid tropics.

Mycorrhiza are classified into three different classes

1. Ectomycorrhiza
2. Endomycorrhiza
3. Ectendomycorrhiza

1. **Ectomycorrhiza:** The fungal symbionts penetrate intercellularly and partially replace the middle lamella between cortical cells of the roots and this hyphal arrangement around such cortical cells is called the `Hartig-net'. The ectomycorrhizal fungal symbionts also form a dense, usually continuous, hyphal network or fungal mantle over the feeder root surface. Normally ectomycorrhizal association takes place on most tree species in the Pinaceae, Salicaceae, Betulaceae, Leguminosae, Ericaceae, Fagaceae and certain other families. The majority of ectimycorrhizal fungi are Basidiomycetes, mainly in the families of Amanitaceae, Tricholomataceae, Rhizopogonaceae and Botetaceae.

2. **Endomycorrhiza:** It penetrates the cortical cells of the feeder roots intracellularly. Such fungal symbionts form large `vesicles' and `arbuscules' in cortical tissues and thus are called `vesicular-arbuscular mycorrhiza'. Such fungi, however, do not form a dense fungal mantle;

instead develop on the root surface a loose, intermittent arrangement of mycelium with large spores. Endomycorrhizal colonisation takes place in most of the agronomic and horticultural crops that do not form ectomycorrhiza. The fungal symbionts are phycomycetes belonging to the family Endogonaceae.

3. **Ectendomycorrhiza:** This intermediate type is present on roots of certain tree species under specific ecological situations. This mycorrhizal type resembles ectomycorrhiza in forming a hartig-net and a fungal mantle, but they also resemble endomycorrhiza in respect of intracellular penetration of cortical tissues by these fungi.

Vesicular-Arbuscular Mycorrhizae (VAM)

Vesicular-arbuscular (VA) mycorrhizal fungi do not induce distinctive alterations of root morphology. Microscopic techniques confirm their presence and/or to determine the extent of infections in the root system.

Structures produced exterior to the root include sporocarps and/or ectocarpic chlamydospores in *Glomus* and *Sclerocystis*, azygospores in *Acaulospora*, and azygospores as well as soil borne vesicles in *Gigaspora*.

VAM infections are characterized by the production of thick-walled inter or intracellular vesicles that are believed to function as endophytic storage organs by the formation of intracellular, haustoria-like arbuscules. The arbuscles are considered as primary structures involved in the bi-directional transfer of nutrients between the fungal symbiont and host plant. *Gigaspora spp.* typically forms only arbuscules. Following penetration of the host epidermis, or in some instances root hairs, hyphae of the fungal symbiont grow inter and/or intracellularly within the cortex. Arbuscules are produced progressively within cells of the inner cortex a short distance behind the advancing hyphal tips. Vesicles produced within infected roots are often thick-walled and otherwise comparable to the chlamydospores produced by the species involved or they may be irregularly shaped and thin-walled, bearing little similarity to chlamydospores. The cytoplasm of vesicles in early stages of development is moderately dense, multinucleated and contains many small lipid droplets as well as glycogen particles. The cytoplasm becomes denser, apparently through condensation, and organelles become more difficult to define ultra-structurally as accumulation of lipids increases during maturation. At maturity, conspicuous lipid globules occupy most of the volume of the vesicle.

Isolation of Arbuscular Mycorrhizal Fungal (AMF) Propagules

Wet sieving and decantation is commonly employed technique for isolation of Arbuscular Mycorrhizal fungal propagules.

Materials

1. Soil samples

1. Glass rod/ magnetic stirrer
2. Sieves
3. Filter papers
4. Funnels
5. Erlenmeyer flasks
6. Root samples
7. Petri dishes
8. Microscope
9. 0.5% Sodium hexametaphosphate
10. Formaldehyde
11. Acetic Acid
12. Alcohol
13. KOH
14. 1% HCl
15. Trypan blue solution

Wet Sieving and Decanting

A. For direct observations of propagules

1. Collect rhizospheric soil (10-30 cm) below the surface and seal the samples tightly in plastic bags.
2. Suspend soil (250 g) in water (1000 ml), stir gently with glass rod/ magnetic stirrer, and allow heavy particles to settle down.
3. Decant the soil sample through 450, 350, 125, 45 and 30 μm sieves stacked vertically.
4. Rinse thoroughly the residue retained on the sieves to remove colloidal matter.
5. Transfer small quantities of residues from sieves to petri dishes for microscopic examination of sporocarps and spores.

B. For further extracting the sieving

1. Sieve soil samples through 2 mm mesh to remove coarse debris and root bits.

2. Add excess water, stir well, and allow heavy particles to sediment.
3. Pour the supernatant through fine sieve (48-38 μm) to retain all arbuscular mycorrhizal spores.
4. Add water to the sediment and repeat the step 2.
5. Rinse thoroughly the material retained on sieve.
6. Transfer the material from the sieve with water into a container.

Direct Observation of Arbuscular Mycorrhizal Spores in Plant Roots

By employing wet sieving-decanting and centrifugation techniques, spores can be isolated and quantified with little information about arbuscular mycorrhizal fungal species occurring on a given plant. Sodium hexametaphosphate, use as soil dispersant, enables direct observation of extrametrical mycelium and spores attached on roots.

1. Immerse roots in 0.5% Sodium hexametaphosphate for 1 minute to clear off the adhering debris.
2. Rinse the roots in tap water and observe roots under microscope for extrametrical hyphae and attached spores.

Quantitative Estimation of Propagules in Soil

Filter paper method is commonly used to enumerate vesicular-arbuscular mycorrhizal (VAM) spores from soil.

Filter Paper Method

1. On an ordinary filter paper disc draw rectangles and number them.
2. Pour spore suspension through this filter paper rested in a funnel and allow the water to drain off.
3. Place the filter paper containing spores on a petri dish and observe under dissecting microscope
4. Count and add spore number from each rectangle, and calculate spore density.

$$\text{Spore density} = \frac{\text{Total number of spores}}{\text{Weight of soil sample in gram}}$$

Mycorrhizal Staining

Trypan blue staining

1. Wash the roots gently to remove adhering soil debris.

2. Preserve the root samples in a mixture of formaldehyde, acetic acid, and alcohol (FAA) in the percentage of 5, 5, and 90 respectively.
3. Immerse the roots in 10% KOH for 60 minutes at 90°C.
4. Acidify the treated roots with 1% HCl for 5 minutes
5. Stain with 0.05% trypan blue at 60°C for 15 minutes. Remove the excess stain with lactoglycerol.

Preparation of Lactoglycerol

Lactic acid - 20 ml

Glycerol - 40 ml

Distilled water - 40 ml

Preparation of Trypan Blue

Trypan blue - 50 mg

Lactoglycerol - 100 ml

References

1. Dubey R.C and Maheshwari D.K. (2004) Practical Microbiology. S. Chand & Company Ltd.
2. Muraleedharan G Nair, Gene R. Safir and Jose O. Siqueira (1991) Isolation and identification of Vesicular-Arbuscular Mycorrhiza stimulatory compounds from clover (*Trifolium repens*) roots 57 (2): 434-39.

Questions

1. Which method is used for direct observation of Arbuscular Mycorrhizal Spores in Plant Roots?
2. How can you enumerate VAM spores from soil?
3. Which reagents are used for preserving roots prior to staining of VAM spores?

16

Test for Antibiotic Sensitivity by Disc Method (Kirby Bauer Method)

Introduction

Antibiotics are the substances produced by one microorganism and are required to inhibit the growth of other microorganisms. Antibiotics are used to combat infectious diseases. The antibiotic sensitivity is very significant due to development of resistance among various microorganisms.

Requirements

1. Swab
2. Nutrient agar
3. Petri plates
4. Bacterial cultures
5. Forceps
6. Antibiotic discs
7. BOD Incubator

Procedure

1. Prepare the bacterial lawn on Nutrient agar plate.
2. Place 5 different antibiotics discs on the lawn by using sterile forceps and incubate at 37°C for 24 hours.

Results and Observations

Examine the plates and measure the zones of inhibition around antibiotic discs. Halo (clear) zones around the discs indicate inhibitory nature of antibiotic.

References

1. Kannan N. (2002) Laboratory Manual in General Microbiology. Panima Publishing Corporation.

2. BensonH. J. (2002). Microbiological applications; a Laboratory Manual in General Microbiology, McGraw Hill, 8th Edition, USA.

Questions

1. What mechanism is responsible for antibiotic resistance?
2. Name two broad spectrum antibiotics.

17

To Study the Germicidal Effect of Ultraviolet Light on Bacterial Growth

Introduction

Most of the bacteria, excepting the photosynthetic bacteria, are damaged by ultraviolet (UV) radiation. UV rays have lower energy content and are capable of producing lethal effect in cells exposed to wavelengths in the range of 200 nm to 300 nm. Nucleic acids absorb these radiations leading to formation of thymine dimers which distort the structure of DNA. Distortion in DNA brings the interferences in DNA replication and transcription. The germicidal effect of UV can be used to determine the minimum amount of exposure required to produce 100% mortality of micro-organisms.

Requirements

1. Nutrient agar medium
2. Card board
3. Incubator
4. Petri plates
5. Inoculating loop
6. Bacterial cultures

Procedure

1. Pour nutrient agar media in sets of four, each in triplicate, marking A,B.C.& D.
2. Streak the specific bacteria in these plates.
3. Remove the lid from these petri plates and cover the half part of each petri plate with black sheet.
4. Allow the petri plates to get exposed for the correct length of time for about 15 minutes, 30 minutes, one hour and two hours from A-D batch, respectively.

5. Covered the plates with lid and incubat at 37^0 C for 48 hours.
6. Record the effect of UV on growth of bacteria.

Results and Observations

Organism	Exposure time and effect of UV			
	15 minutes (A)	30 minutes(B)	1 hour(C)	2 hours(D)
Bacteria 1				
Bacteria 2				

Reference

Kocher G.S. (2012). Practical Teachings in Microbiology. Narendra Publishing House, New Delhi (India)

Questions

1. What is the effect of UV radiations on survival of bacteria?
2. What is the average exposure time of UV radiations to cause inactivation of bacteria?

18

To Study the Effect of Temperature on Growth of Microorganisms

Introduction

The most important factor that influences the activity of bacterial enzymes is temperature. Most of the organisms grow under mesophilic conditions (at25°C). If the optimum temperature is increased, the rate of growth of organism is affected due to changes in protein structure which results denaturation of critical cell protein and ultimately inhibitsthe growth. If temperature is below critical temperature, then there is complete stoppage of growth as less activation energy is available for critical process to occur. However, there are certain organisms which can tolerate extreme temperatures viz.

1. Psychrophiles: Tolerate and grow at < 20°C temperature
2. Mesophiles: Tolerate and grow at between 20°C and 45°C temperature
3. Thermophiles: Tolerate and grow at >45°C temperature.

Requirements

1. Erlenmeyer flasks (100 ml)
2. Culture of *E. coli*
3. Nutrient broth
4. Pipettes
5. Spectrophotometer

Procedure

1. Inoculate 10^6 number of cells per ml of *E. coli* in 100 ml conical flasks with 50 ml nutrient broth. The experiment is performed in the set of three.
2. Incubate these flasks at 4° C, 20° C, 25° C, 37° C and 45° C respectively on a rotary shaker for 24 hours.
3. Observe the turbidity with the help of spectrophotometer after calibration at 550nm optical density. The turbidity in the tubes is compared with control.

Results and Observations

Plot a graph between temperature (X-axis) and optical density at 550 nm (Y-axis). The crest of bell-shaped curve will indicate optimum growth temperature.

References

1. Dubey R.C and Maheshwari D.K. (2004) Practical Microbiology. S. Chand & Company Ltd.
2. Case C.L. and Johnson T.R. (1984). Laboratory Experiments in Microbiology. The Benjamin Cummings Publishing Co., California, USA.

Questions

1. Why do different microorganisms grow at different temperatures?
2. What advantages thermophilic bacteria have in growing at higher temperature?

19

Effect of Osmotic Pressure on Growth of Bacteria

Introduction

The water potential plays an important role on growth of bacteria. If the medium has higher concentration of solute, it is said to be hypertonic. Due to this, the cytoplasm dehydrates and shrinks, resulting in inhibition of growth of bacteria. If solute concentration is less, the medium is hypotonic and hence maximum osmotic pressure on the cell.

Requirements

1. Cultures of *E.coli* and *Bacillus*
2. Nutrient agar
3. Sodium Chloride (stock solution at 5 M concentration)
4. Inoculating loop
5. BOD incubator

Procedure

1. Prepare five nutrient agar plates with different concentrations of NaCl (0.5%, 1%, 5%, 10% and 15%w/v) each in triplicates.
2. Draw two lines at the bottom of petri plates and streak *E. coli* and *Bacillus* on respective half section of plate.
3. Incubate the petri plates at 37°C for 48 hours.

Results and Observations

Observe the rate of inhibition with the increase in salt concentration and compare it with control.

Reference

Kocher G.H. (2012) Practical teachings in Microbiology. Narendra Publishing House, New Delhi (India)

Questions

1. Why Pickles and Jams have high concentration of salt and sugar, respectively?
2. Why the osmotic pressure requirements of bacteria and fungi are different?
3. What kinds of bacteria can be expected in sea water?
4. Differentiate between hypertonic, hypotonic and isotonic solutions and their effects on cells?

20

Effect of pH on Growth of Bacteria

Introduction

The pH (hydrogen ion concentration) of the environment provides the maximum effect on the growth of bacteria. Each organism grows at its optimal pH. Due to variation in other factors such as temperature, osmotic pressure etc marked changes in pH have also been observed.

Requirements

1. Erlenmeyer flasks (100 ml)
2. Culture of *E. coli*
3. Pipettes
4. Spectrophotometer
5. pH meter

Procedure

1. Inoculate 10^6 number of cells per ml of *E. coli* in 100 ml conical flasks with 50 ml nutrient broth of varying pH values i.e. 5, 6, 7, 8, 9 and 10. The experiment is performed in the set of three.
2. Incubate these flasks at 37°C temperature for 48 hours.
3. Observe the turbidity at various intervals (4 hours) with the help of spectrophotometer after calibration at 550nm optical density. The turbidity in the tubes is compared with control.
4. Draw a graph between incubation time (X-axis) and pH (Y-axis).

Results and Observations

Maximum Optical density along Y-axis will provide the optimum pH for the growth of a particular culture.

References

Benson, H.J. (2002). Microbiological applications; a Laboratory Manual in General Microbiology, McGraw Hill, 8th Edition, USA.

Questions

1. Why bacteria do not survive under acidic conditions?
2. Why Fungi prefer alkaline conditions for their growth?

21

Measurement of Phosphatase Activity of Soils

Introduction

Various organic orthophosphoric mono-and di-esters in soils are hydrolyzed when incubated with toluene-treated or irradiated soils, indicative of phosphatase activity. Assays have been identified, based on the rates of release of orthophosphate. Phosphatase activities of field soils showed little or no variation with season but were influenced by the cropping system and nature of the plant cover. Activities generally increased after the addition of inorganic and organic fertilizers to soils. However, inhibition of phosphatase by phosphatic fertilizers and by liming of soils (application of ca and Mg) has been reported. Phosphatase activities were directly related to the organic-P of soils but were not consistently related to available-P contents.

Phosphatases have been used to describe a broad group of enzymes that catalyze the hydrolysis of both esters and anhydrides. These are classified into five categories

1. Phosphoric monoester hydrolase
2. Phosphoric diester hydrolase
3. Triphosphoric monoester hydrolases
4. Enzymes acting on phosphoryl containing anhydrides
5. Enzymes acting on P-N bonds such as the phosphoamidase.

Phosphomonoesterase: Acid phosphatase and alkaline phosphatase orthophosphoric monoester phosphohydrolase have been studied extensively. The enzymes are classified as acid and alkaline phosphatase because they show optimum activities in acid and alkaline ranges, respectively. Because of the importance of these enzymes in soil organic-P mineralization and plant nutrition considerable literature has accumulated on Phosphomonoesterase in soil. These enzymes are known to hydrolyze, a variety of Phosphomonoesters. The hydrolysis of glycerophosphate, phenylphosphate, naphthyl phosphate, and p-nitrophenyl phosphate have also been reported. The inverse relationship between phosphatase activity and soil pH suggests that either the rate of synthesis and release of this enzyme by soil micro-organisms or stability of this enzyme is related to soil pH. Since higher plants are deprived of alkaline phosphatase activity, it seems they are derived from microorganisms. The apparent Km value for acid phosphatase in soils range from 1.3 to 4.5 mM and those for alkaline phosphates range from 0.4 to 4.9 mM. Adsorption of phosphatase one-day mineral significantly alters their kinetic parameters. Because of the importance of these enzymes in soil organic Phosphorous mineralization and plant nutrition, considerable literature has accumulated on phosphomonoesterase in soils. The general equation of the reaction catalyzed by acid and alkaline phosphatase is:

$$\text{R-O-}\overset{\overset{\text{O}}{\|}}{\underset{\underset{\text{O}^-}{|}}{\text{P}}}\text{-O- + H}_2\text{OR.OH+HO-}\overset{\overset{\text{O}}{\|}}{\underset{\underset{\text{O}^-}{|}}{\text{P}}}\text{-O}$$

Several methods have been proposed for estimation of phosphomonoesterase activities in soils. The basic difference is in the substrate used and consequently in the technique employed in measuring the product of hydrolysis of the substrate by phosphatase enzymes.

Of the various methods available for assay of phosphatase activity in soils, the method developed by Tabatabai and Bremner (1969) is described here as considered the most rapid, accurate, and precise one. It involves colorimetric estimation of the p-nitrophenol released when soil is incubated with buffered sodium p-nitrophenyl phosphate solution and toluene.

Principle

The procedures described for assay of phosphomonoesterase activities are based on calorimetric estimation of the p-nitrophenol released by phosphatase activity when soil is incubated with buffered (pH 6.5 for acid phosphatase activity and pH 11.0 for alkaline phosphate activity) sodium p-nitrophenyl phosphate solution and toluene. The calorimetric procedure used for estimation of p-nitrophenol is based on the fact that alkaline solutions of this phenol have a yellow colour (acid solutions of p-nitrophenol and acid and alkaline solutions of p-nitrophenyl phosphate are colourless). The intensity of yellow colour is measured at 420 μm wavelength. The $CaCl_2$-NaOH treatment described for extraction of p-nitrophenol after incubation for assay of acid and alkaline phosphatase serves

i) to stop phosphatase activity,

ii) to develop the yellow color used to estimate this phenol, and

iii) to give quantitative recovery of p-nitrophenol from soils.

Materials

Incubation flasks

Incubator

Colorimeter or spectrophotometer.

Toluene

Modified universal buffer (MUB) stock solution: Dissolve 12.1 g of tris hydroxymethyl aminomethane (THAM), 11.6 g of maleic acid, 14.0 g of citric acid and 6.3 g of boric acid (H_2BO_3) in 488 mL of 1N sodium hydroxide (NaOH) and dilute the solution to 1 L with water. Store it in a refrigerator.

Modified universal buffer, pH 6.5 and 11: Place 200 mL of MUB stock solution in a 500 mL beaker containing a magnetic stirring bar and place the beaker on a magnetic stirrer. Titrate the solution to pH 6.5 with 0.1 M hydrochloric acid (HCl), and adjust the volume to 1 L with water. Titrate another 200 mL of the MUB stock solution to pH 11 by using 0.1 M NaOH and adjust the volume to 1 L with water.

p-nitrophenyl phosphate solution, 0.05 M: Dissolve 0.840 g of disodium p-nitrophenyl phosphate tetrahydrate in about 40 mL of MUB pH 6.5 (for assay of acid phosphatase) or pH 11 (for assay of alkaline phosphatase) and dilute the solution to 50 mL with MUB of the same pH. Store the solution in a refrigerator.

Calcium chloride ($CaCl_2$), 0.5 M: Dissolve 73.5 g of $CaCl_2.2H_2O$ in about 700 mL of water, and dilute the volume to 1 L with water.

Sodium hydroxide (NaOH), 0.5 M: Dissolve 20 g of NaOH in about 700 mL of water and dilute the volume to 1 L with water.

Standard p-nitrophenol solution: Dissolve 1 g of p-nitrophenol in about 700 mL of water, and dilute the solution to 1 L with water. Store the solution in a refrigerator.

Procedure

1. Place 1 g of soil (< 2 mm) in a 50 ml Erlenmeyer flask.
2. Add 0.2 ml of toluene, 4 ml of MUB (pH 6.5 for assay of acid phosphatase or pH 11 for assay of alkaline phosphatase) and 1 ml of p-nitrophenyl phosphate solution made in the same buffer.
3. Swirl the flask for a few seconds to mix the contents. Stopper the flask, and place it in an incubator at 37°C.
4. After 1 hour, remove the stopper; add 1 ml of 0.5 M $CaCl_2$ and 4 ml of 0.5 M NaOH.

5. Again swirl the flask for a few seconds and filter the soil suspension through a Whatman no.1 folded filter paper. Measure the yellow color intensity of the filtrate with a colorimeter.
6. Calculate the p-nitrophenol content of the filtrate by reference to a calibration graph plotted from the results obtained with standards containing 0, 10, 20, 30, 40 and 50 μg of p-nitrophenol.
7. To prepare this graph, dilute 1 ml of the standard p-nitrophenol solution to 100 ml in a volumetric flask and mix the solution thoroughly.
8. Then pipette 0, 1, 2, 3, 4, and 5 ml aliquots of this diluted standard solution into 50 ml Erlenmeyer flasks.
9. Adjust the volume to 50 ml by addition of water and proceed as described for p-nitro phenol analysis of the incubated soil sample (i.e., add 1 ml of 0.5 M $CaCl_2$ and 4 ml of 0.5 M NaOH, mix and filter the resultant suspension).
10. If the colour intensity of the filtrate exceeds that of 50 μg of the p-nitrophenol standard, an aliquot of the filtrate should be diluted with water until the colorimeter reading falls within the limits of the calibration graph.
11. Controls should be performed with each soil analyzed to allow for colour not derived from p-nitrophenol released by phosphatase activity. To perform controls, follow the procedure described for assay of phosphatase activity, but make the addition of 1 ml of p-nitrophenol solution after the additions of 0.5M $CaCl_2$ and 4 ml of 0.5M NaOH (i.e. immediately before filtration of the soil suspension).

Calculations

Phosphatase activity (μg /g/hour)= C /weight of soil x time

C = Amount of PNP produced (μg) as read from standard curve

References

1. Tabatabai, M.A. and Bremner, J.M. (1969) Use of p-nitrophenyl phosphate for assay of soil phosphatase activity. *Soil Biol. Biochem*. 1: 301-307.
2. Tarafdar, J.C and Chhonkar, P.K. (1978) Status of phosphatases in the root-soil interface of leguminous and non-leguminous crops. *Z. Pflanzenirnaehr Bodenkund.* 14: 347-351.

Questions

1. What is the function of toluene in phosphatase activity?
2. Phosphatase enzyme acts on which substrate for the production of yellow colour?
3. At which wavelengh the maximum absorption of yellow colour produced by p.nitrophenol takes place?

22

Determination of Urease Activity of Soils

Introduction

Urea is one of the common fertilizers used as a source of nitrogen. It has nitrogen in amide form ($-NH_2$) and this form of nitrogen is not utilized by plants. So it has to be hydrolysed. The enzyme, urease hydrolyses the urea to ammonia and ammoniacal nitrogen is taken up by plants or further nitrified.

$$CO(NH_2)_2 + H_2O \longrightarrow CO_2 + 2NH_3$$

The measurement of activity of this enzyme in soil is important from soil fertility point of view. There are two methods for determination of the urease activity of soils.

- First involves the measurement of end products of urea hydrolysis. i.e., NH_3 or CO_2.
- Second method involves the colorimetric measurement of urea remaining in soil after hydrolysis. This method is common in use.

Principle

A known amount of urea is added to soil. The enzyme, urease will hydrolyse it. Urea left in soil after incubation at 37°C for 5 hours is determined colorimetrically. The amount of urea hydrolysed per gram of soil per 5 hours is estimated from the difference between initial amount of urea added and that recovered after the incubation period. It is expressed as amount of urea hydrolysed per gram of soil per minute, which represents the urease activity of soil.

Materials

1. Conical flasks (150 ml)
2. Incubator
3. Volumetric flasks (50 ml.)
4. Water bath (Temperature controlled)
5. Balance
6. Spectrophotometer

Preparations of Solutions: Urea solution (2%w/v): Add 2 grams of urea (AR grade) in 1000 ml of distilled water.

1. Phenylmercuric acetate (PMA) solution: Dissolve 50 mg of PMA in one litre of distilled water.
2. Potassium chloride-PMA solution (2M KCl- PMA): Dissolve 150 grams KCl (Reagent grade) in 900 ml of distilled water, and then add 100 ml of PMA solution to make 1000 ml.
3. Diacetyl Monoxime solution (DAM): Dissolve 2.5 grams of DAM in 100 ml. of distilled water.
4. Acid reagent: Add 300 ml of 85% phosphoric acid and 10% (10 ml of concentrated H_2SO_4 to 100 ml of distilled water) and make the volume to 500 ml with distilled water.
5. Thiosemicarbazide (TSC): Dissolve 0.25gram TSC in 100 ml of distilled water.
6. Colouring reagent: It is prepared immediately before use by adding 25 ml of DAM solution and 5 ml of TSC solution to 250 ml of acid reagent.
7. Standard urea solution (for preparation of standard curve).

   Dissolve 250mg of urea in about 700 ml 2M KCl-PMA solution and dilute to one litre

Procedure

1. Grind the soil sample and sieve it (< 2mm).
2. Take 5.0 g of this sample in 150 ml volumetric flask.
3. Add 5.0 ml of 2000 ppm urea solution.
4. Make the moisture at 50% of water holding capacity.
5. Incubate the flask at 37°C for 5 hours in an incubator.
6. Add 50 ml of 2M KCl - PMA solution and shake for 60 minutes on a shaker and filter.
7. Take 1 ml of this extract in 50 ml volumetric flask. Add 10 ml of extracting reagent and 30 ml of colouring reagent.
8. Boil it in a boiling water bath for 30 minutes. Heat it in an oven at 100°C for 20 minutes.
9. Cool at room temperature for 30 minutes.
10. Make the volume with distilled water, mix thoroughly and measure the intensity of red colour at 527 nm wavelength on Spectrophotometer.
11. Also prepare standard curve using 0, 25, 50, 100, 150, 200 μg/L urea concentration. Dilute 10 ml of the standard urea stock solution to 100 ml

with 2M KCl-PMA solution in a volumetric flask and mix thoroughly. Then pipette 0, 1, 2, 4, 6 and 8 ml of aliquots of this working standard solution into 50 ml volumetric flasks. Adjust the volume to 10 ml by adding 2M KCl-PMA solution and proceed as described for urea analysis of the soil extracts.

Observations and Calculations

1.	Weight of soil sample taken	= 5.0 g
2.	Concentration of urea added in soil	= 2000 μg/g soil
3.	Volume of extract added in soil	= 50 ml.
4.	Volume of extract taken for colour development	= 1.0 ml.
5.	% T	= A
6.	Concentration of urea as read from standard curve against A value	= B μg
7.	Total urea extracted from the soil	= 50xB μg
8.	Amount of urea remaining in the soil	= 50xB/5= C μg/g
9.	Amount of urea hydrolysed	= 2000- C= D μg
10.	Urease activity (μg urea /g dry soil /min)	= D /300

References

1. Bremner J.M. and Douglas L.A. (1971) Inhibition of urease activity in soils. *Soil Biol. Biochem.* **3**: 297-307.
2. Pal S. and Chhonkar P.K. (1981) Urease activity in relation to soil characteristics. *Pedobiologia*, 21: 152-58.

Questions

1. Which extracting reagent is used for determining urease activity?
2. Which products are being produced due to urea hydrolysis and give red colouration?

23

Measurement of Dehydrogenase Activity of Soils

Introduction

All biological reactions in soil are catalyzed by enzymes. Soil enzyme activities are believed to indicate the extent of specific processes in soil and in some cases, can act as indicators of soil fertility. Biological oxidation of organic compounds is generally a dehydrogenation process and there are many dehydrogenases that carry out the following reaction:

$$XH_2 + A \longrightarrow X + AH_2$$

XH_2 is an organic compound and is a hydrogen donor and A is a hydrogen acceptor. These systems are integral part of the microorganisms, so dehydrogenase activity can be correlated with active population.

Principle

Measurement of dehydrogenase activity involves determination of Triphenyl formazan (TPF) produced by the reduction of Triphenyl tetrazolium chloride (TTC). TTC is a redox indicator, a yellow coloured water soluble salt which possesses the property of being easily transformed into, methanol soluble formazan (intensely coloured, water insoluble) by reduction. The apparent redox potential of TTC is about -0.08V, which makes this compound act as an electron acceptor for many dehydrogenases. TPF formed is extracted with methanol. The general reaction involved is:

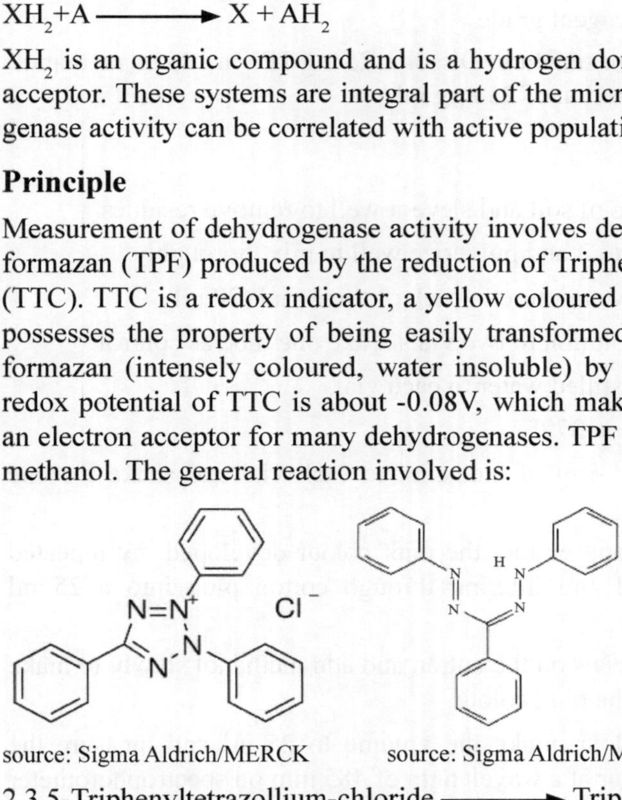

$HN\ N{=}\ CN = N + H^{+}Cl^{-1}$

source: Sigma Aldrich/MERCK   source: Sigma Aldrich/MERCK

2,3,5-Triphenyltetrazollium-chloride ⟶ Triphenyl formazan

Materials

1. Fresh soil
2. Vials (20 ml)
3. Funnels
4. Cotton
5. Pipette (graduated, 5.0 ml)
6. Volumetric flasks (25 ml)
7. Distilled water
8. Electronic balance
9. Incubator
10. Spectrophotometer
11. 2, 3, 5 - Triphenyl tetrazolium chloride (3%). Weigh 3 gm of TTC and dissolve it to 100 ml of distilled water.
12. Methanol, analytical grade
13. Calcium carbonate, reagent grade
14. Triphenyl formazan for standard: Dissolve 25 mg TPF in 100 ml methanol. It will give a concentration of 250 µg ml^{-1}

Procedure

1 Weigh exact 10 grams of soil and sieve it well to remove residues.
2. Mix 0.1gram $CaCO_3$ to it and pulverize well to mix thoroughly.
3. Place 3 grams of above mixture in each of three vial (20 ml).
4. Add 0.5 ml of TTC solution to two vials. Take one tube as control.
5. Also add 1.5 ml of distilled water to each vial.
6. Incubate for 24 hours at 37°C.
7. After incubation, add 5ml of methanol to each vial and shake for one minute.
8. Unstopper the vials and extract the pink colour developed, by repeated washings to the soil and filtering through cotton plug into a 25 ml volumetric flask.
9. Empty the soil from vials on the cotton and add methanol slowly to make the cotton free from the pink colour.
10. After removal of colour, make the volume to 25 ml and measure the intensity of pink colour at a wavelength of 485 mm on spectrophotometer or colorimeter using methanol as blank.

For standard curve, take 0, 0.5, 1.0, 1.5, and 2.0 ml on the above made TPF solution into 25 ml flask and make volume with methanol. It will give as the solution of 0, 125, 250, 375, 500 μg TPF concentration. Measure the intensity same ways for sample.

Observations and Calculations

Weight of Soil taken = 3grams

Readings of standard curve

Concentrations (μg)	% T
0	
125	
250	
375	
500	

Per cent Transmittance of the Sample =

Amount of Formazan (μg) produced read from the standard curve = C μg

Dehydrogenase Activity = C/3x24 = D μg TPF g^{-1} soil hr^{-1}

Reference

Klein D.A., Loh T.C. and Goulding R.L. (1971) A rapid procedure to evaluate dehydrogenase activity of soils low in organic matter. *Soil Biol. Biochem*. 3: 385-87.

Questions

1. Why triphenytetrazolium chloride is used in dehydrogenase assay?
2. What is the nature of triphenytetrazolium chloride and triphenyl formazan?
3. What is the function of calcium carbonate?
4. At which wavelengh pink colour of TPF shows maximum absorption?

24

Nitrogenase Activity of Legume Nodule by Acetylene Reduction Assay (ARA)

Introduction

The enzyme nitrogenase is produced in all N_2-fixing systems. The activity of nitrogenase enzyme relates to the biological nitrogen fixation, which can be determined directly by measurement of fixation product, ammonia and the further products of its assimilation. The activity can also be measured indirectly by measuring the ability of nitrogenase to reduce the number of substrates other than N_2. Substrates such as nitrous oxide (N_2O), azide ion (N_3^-), cyanide (CN^-) and isocyanide (NC^-) are reduced and the requirements for ATP and reductant are similar to those of N_2 fixation. The reduction of protons to molecular hydrogen has been used to measure the nitrogenase activity.

Dilworth (1966) and Schollhorn and Burris (1966&1967) observed the inhibitory effect of Acetylene (C_2H_2) on N_2 fixation by extracts of *Clostridium pasteurianum*. Dilworth (1966) further observed by mass spectrometry that the C_2H_2 was reduced to C_2H_4. Both gases are readily measured at very low concentrations by gas chromatographic techniques. As a consequence of its high sensitivity, low equipment and resource costs, simplicity and rapidity of use, the acetylene reduction assay quickly became a standard procedure in many laboratories.

The only drawback is that it is a onetime measurement and fails to integrate the total nitrogen fixed in a system. However, to compare the treatments of nitrogen fixing systems, this technique remains the best.

Principle

The method has been used extensively to estimate nitrogenase activity and a theoretical conversion factor of 4 mole of C_2H_4 produced to 1 mole of N_2 reduced is more commonly used these days. Actually, conversion factors range between 1.5 to 25 and must be determined experimentally.

For nodulated plants, measurable acetylene reduction activity occurs for a single effective nodule almost immediately on exposure to C_2H_2. The quantity of C_2H_2 reduced in a unit time by a unit mass of nodules is measured by gas chromatography using mixture of pure ethylene as the standard. The closed system that is described is useful for determining nitrogenase activity, if nodules are active in N_2 fixation

Materials

1. Calcium carbide (CaC_2)
2. Ethylene for preparation of standards.
3. Apparatus for generating C_2H_2
4. Conical flask 100 ml capacity.
5. Rubber septa, 5.5 mm diameter.
6. Plastic syringes, 50 mL
7. Plastic syringes, 5 mL
8. Plastic syringes, 1 mL
9. Nodulated roots
10. Gas chromatograph equipped with H flame ionization detector
11. Porapak Q column, 1.5 m long and 3 mm diameter
12. Nitrogen carrier gas.
13. Hydrogen fuel gas.

Procedure

1. Generate C_2H_2 from CaC_2 by adding it to the water in the apparatus. The apparatus allows collection of the generated C_2H_2 in a reservoir created by the displacement of water. Acetylene may be partially purified by passing the gas through concentrated sulphuric acid and water.
2. Intact nodules are used to determine the nitrogenase activity. The plants are uprooted and the root system with intact nodules is separated from the shoot system and is placed in a 60 ml vial provided with serum stopper.
3. Replace 10% of air with acetylene gas and incubate the vial for 60 minutes a 30°C.
4. Inject 0.1 ml of acetylene gas in GLC septum. The other conditions are:

Oven temperature	80°C
Flame ionization detector (FID) temperature	110°C
Carrier gas (N_2)	25-30 ml/min
Fuel gas (H_2)	25-30 ml/min

5. Measure the peaks of acetylene and ethylene on detector.

6. Inject the gas mixture from vials @ 0.1 ml under the same conditions as given in 4 and measure the ethylene peak obtained and compare with standard.
7. Now remove nodules from roots and take their weight.

Observations and Calculations

Before using the gas chromatograph for assays, it is necessary to check the linearity of the response over a range of concentrations of both C_2H_4 and C_2H_2. Subsequently one or two different standards should be run each day, provided the response is linear. Changes in detector response to a given concentration may occur due to detector fouling.

The nitrogenase activity of the sample was determined as:

µM C_2H_2 reduced h^{-1} plant wt^{-1} = CU x K x V

CU = Chart Units

V = volume of the vial used

K = µM C_2H_2 reduced per chart unit. K is calculated as 4.46/x, where x is the position where peak is obtained and 4.46 is the volume of 0.1 ml of any gas at NTP.

Reference

Kocher G.H. (2012) Practical teachings in Microbiology. Narendra Publishing House, New Delhi (India)

Questions

1. Name the carrier and fuel gases used for determination of nitrogenase activity?
2. Which column is used for Gas liquid chromatography?
3. Which conditions are required for Gas liquid chromatography?

25

Measurement of Methane using Gas Chromatograph

Introduction

The major biogenic gases, which contribute to global warming, are carbon dioxide, methane and nitrous oxide. Out of these, methane is the most important gas because its thermal absorption is 30 times greater than carbon dioxide, though its atmospheric concentration is about 200 times less than carbon dioxide. Atmospheric measurements of greenhouse gases indicate that methane concentration had increased by 1 per cent annually to 1.7 ppm (by volume) during the last decade. The steady increase of atmospheric methane concentration in the past 200 years has attracted much attention.

Flooded rice soils provide an optimum environment for methane production and emission. Flooding a soil creates essential redox potential, anaerobic decomposition of organic matter and stabilizes the soil pH near neutral, thus makes the congenial environment for methane emission.

Rice is the world's most important food crop. It provides more than half of the daily food for one of the every three persons on the earth. Thus, we cannot quit growing rice to reduce methane emission. Moreover, projected global population levels indicate that demand of rice will increase to 760 million tons year^{1} by 2020. It is most likely to be met by the existing cultivable wetland area through intensifying rice production, which will result into greater methane emission.

Formation of biogenic methane is due to the activities of methanogenic bacteria that can metabolise in the absence of oxygen. These bacteria possess an co-enzyme known as F-420 hydrogenase, which is sensitive to oxygen. In rice fields methanogens occur in the reduced layer of soil.

Twenty genera of methane producing bacteria have been described by Garcia (1990) but only a few including *Methanobacteria* and *Methanosarcina* have been isolated from rice soil while *Methanospirillum* and *Methanocorpusculum* were isolated from fresh water sediments.

Methanogens are distributed in natural environments on the basis of their adaptation to temperature, pH and salinity ranges. Most of the methanogens are mesophilic with optimum temperature of 30-40^{0}C but some are thermophilic (<40^{0}C). Generally, methanogens are neutrophilic with an optimum range of pH 6-8. A few are alkaliphilic, having optimum growth at pH 8-9. No acidophilic strains have been isolated.

On the basis of substrate, these are also grouped as hydrogenotrophic methanogens, which oxidize hydrogen and reduce carbon dioxide to form methane. Acetotrophic methanogens utilize acetate. The growth of virtually all the methanogens is stimulated by acetate. Methylotrophic methanogens utilize methylamines.

Sixty per cent of hydrogenotrophic species also use formate ion. Few species use hydrogen to reduce methanol (hydrogen methylotrophic methanogens) other form methane in the presence of carbon dioxide and alcohol as H donor (alcohol trophic methanogens). Four types of bacteria are needed to produce methane:

a) Hydrolytic and fermenting bacteria,
b) Hydrogen reducing bacteria,
c) Homoacetogenic bacteria,
d) Methanogenic bacteria.

The first group hydrolyses polymers and ferments the resulting monomer to smaller molecules and carbon dioxide. Methanogens can convert these to methane from formate ion, acetate and a few others simple compounds. Fermentation products like fatty acids, alcohols, aromatic and others cannot directly be utilized. They are first oxidized by obligate hydrogen producing bacteria to carbon dioxide and acetate.

Methane production occurs only in the anaerobic ecosystem. The strictly anaerobic methanogenic bacteria are responsible for the biogenic methane production. They belong to Archaebacteria and can be subdivided according to their substrate requirement for methanogenesis.

1. Obligate-chemo lithotrophic methanogens, use exclusively carbon dioxide and hydrogen for methane generation.

   $$CO_2 + 4\,H_2 \longrightarrow CH_4 + 2H_2O$$

2. Quasi-chemo lithotrophic methanogenic bacteria, which are able to use H_2, HCOOH or CO besides CO_2

   $$4HCOOH \longrightarrow CH_4 + 3CO_2 + 2H_2O$$

   $$4CO + 2H_2O \longrightarrow CH_4 + 3CO_2$$

3. Methylotrophic methanogenic bacteria are able to use methyl group containing compounds like methanol, acetate, and trimethylamine as substrate for methane generation.

$$CH_3COOH \longrightarrow CH_4 + CO_2$$

$$4CH_3OH \longrightarrow 3CH_4 + CO_2 + 2H_2O$$

$$4(CH_3)_3\text{-}N + 6H_2O \longrightarrow 9CH_4 + 3CO_2 + 4NH_3$$

Methane is inert under anaerobic condition and of no use for microbes and hence emitted.

Materials

1. GLC fitted with Parapak T column and FID detector
2. Incubation Flasks
3. Methane (CH_4) collecting chamber
4. Gas tight syringes
5. Methane Gas standard

Preparation of pure methane

Reaction of sodium acetate with soda lime fused or anhydrous sodium acetate is mixed with about three times its weight of soda lime. The mixture is heated in hard glass test tube and methane gas is collected by the displacement of water:

$$CH_3COONa + NaOH \ ^{CaO\ \Delta}\ CH_4 + Na_2CO_3$$

Procedure

Collection of Gas: Suitable size chambers are installed in the field with help of a base plate. The gas samples are collected in the gas tight syringes as and when required. The Chamber is made of Plexiglass and is measuring 34.3cm x 53.2cm x72.4 cm.

Analysis: Gas chromatograph equipped with flame ionisation detector is needed to detect methane. Flow rates of N_2, H_2 and air are kept at 20 ml minute^{-1}, 25 ml minute^{-1}, and 235 ml minute^{-1}, respectively. The temperatures for oven, detector and injector are kept at 80°C, 120°C and 100°C, respectively. A glass column of 2-meter length with internal diameter 1/4" and thickness of 2 mm containing Parapak -T (80-100 mesh) is used to separate methane in the samples. The GLC is calibrated with methane standard. The chromatograms can be processed by the integrator. Unknown samples are injected into the Gas chromatographs and results are processed by the integrator.

Observations and Calculations

Before using the gas chromatograph for assays, it is necessary to check the linearity of the response over a range of concentrations of CH_4. Subsequently one or two different standards should be run each day, provided the response is linear. Changes in detector response to a given concentration may occur due to detector fouling.

Methane flux was calculated in mg/m²/hr. =

$$= \frac{(.\text{Concentration (ppm) x107 x 16 x EBV x60})}{(10^6\text{x 22400 x time (in Min) x 34.3 x53.2})}$$

. Concentration (ppm) = (Peak area of sample / (Peak Area of Standard) x concentration of Standard (1.95ppm)

Dimensions of the Chamber

Height of Chamber = 72.4 cm

Width of Chamber = 34.3 cm

Length of Chamber = 53.2 cm

EBV (Effective box Volume) cm^3

= (72.4-water depth) X 34.3 X 53.2-Total biomass volume

EBV at Standard temperature and pressure

$$= \frac{\text{273 X barometric pressure X EBV}}{\text{760 X (Temperature + 273)}}$$

Questions

1. How methane-producing bacteria come under diverse class?
2. Give the food requirements of methanogenic bacteria.

26

Determination of Total Nitrogen in *Azotobacter* Cells

Introduction

Demonstration of biological nitrogen fixation is usually accomplished by estimating the total nitrogen through use of some modification of the Kjeldahl procedure. This technique is accurate and reliable whenever a medium initially low in nitrogen is employed.

Principle

The dried organism is digested with sulphuric acid. Organic and nitrate nitrogen is converted to ammonium sulphate and the ammonia gas is distilled into boric acid forming ammonium borate making weak boric acid alkaline. The ammonium borate is back titrated with standard sulphuric acid until the reaction mixture changes its colour from green to pink due to formation of ammonium sulphate. The standard acid used in back titration of ammonium borate is a measure of nitrogen content of the organism.

Materials

1. Pure *Azotobacter* cells
2. Jensen broth (Nitrogen free broth)
3. Digestion tube
4. Distillation unit
5. Top burette
6. Flasks
7. Digestion mixture ($CuSO_4$: K_2SO_4: Selenium powder (20:3:1) + 500 g Na_2CO_3)
8. 4% Boric acid in distilled water (add 20 ml of mixed indicator to 1 litre of 4% Boric acid till the pH reaches 4.5-5.1)
9. NaOH (40%)
10. No, 35 H_2SO_4 detecter tubes

Procedure

1. Pure culture of *Azotobacter* is raised on Jensen agar medium.
2. 10^6 *Azotobacter* cells /ml are inoculated in 100 ml Jensen broth under shake at 28± 2°C for 5 days.
3. Uninoculated broth served as control.
4. The inoculated organism is dried and is digested with 30 ml of concentrated H_2SO_4 and 1.2 g of digestion mixture till clear.
5. The digested contents were cooled and volume made to 25 ml in volumetric flask with distilled water.
6. This solution was distilled in Micro kjeldahl distillation apparatus by adding 20 ml of 40% NaOH.
7. The liberated ammonia was trapped in 5 ml boric acid with mixed indicator.
8. About 50 ml of green distillate collected in 50 ml conical flask was titrated with N/35 H_2SO_4 till the distillate changed from green to faint pink. Uninoculated control was also run simultaneously.

The N content was calculated by using the following equation

$$\% \text{ N in sample} = \frac{(\text{a-b}) \times 0.00014 \times 100}{35}$$

Reference

Kocher G.H. (2012) Practical teachings in Microbiology. Narendra Publishing House, New Delhi (India)

Questions

1. Why Nitrogen free broth is used for determination of total Nitrogen in *Azotobacter* cells?
2. What is the function of Boric acid?

27

Measurement of Microbial Biomass Carbon in Soils by Fumigation-Extraction Technique

Introduction

The soil microbial biomass is an important component of the soil organic matter that regulates the transformation and storage of nutrients. It is a labile component of the soil organic fraction containing 1 to 3% of the soil carbon and up to 5% of the total soil N. The microbiologically mediated processes affect ecosystem functions associated with nutrient cycling, soil fertility, global C change and soil organic matter turnover. The size and activity of the soil microbial biomass must be assessed to fully understand nutrient flux in managed and natural ecosystems.

Soil microbial biomass estimations are useful in investigations that compare temporal nutrient fluctuations along natural and perturbed gradients. The effect of tillage, crop rotations and soil type on organic carbon and nutrient turnover can be assessed by following nutrient pools and activity associated with the soil microbial biomass. Microbial biomass has been shown to be sensitive indicator of differences in sustainable cropping system. The toxicity of pollutants and the degradation of organic compounds can be monitored by following changes in soil microbial biomass.

Principle

The effect of fumigants on the metabolism was established early during this century. The respiration rate of a fumigated soil is initially less than unfumigated sample but as time proceeds, the respiration rate of fumigated soil exceeds that of the unfumigated sample and essentially subsides to a lower level. The temporary flush of CO_2 from the fumigated soil is primarily due to the decomposition of microbial components from the lysed microorganisms. In addition, an increase in the NH_4^+ pool occurs as a result of the mineralization of nitrogenous substrates from the lysed microorganisms. The increase in CO_2 evolution and extractable NH_4^+ from fumigated samples has been used to estimate the size of soil microbial biomass.

The living soil biomass is killed by fumigation with chloroform vapour. Afterwards the soil is extracted with potassium sulphate solution. The organic carbon in the filtrate is determined and converted to biomass Carbon.

Materials

1. 50 ml capacity beakers
2. 0.5 M Potassium sulphate solution: Dissolve 87.12 g of Potassium sulphate in distilled water and make up the volume to 1000 ml in a volumetric flask.
3. 33.3 mM Ferrous ammonium sulphate solution: Dissolve 13.07 g of ferrous ammonium sulphate in distilled water and make the volume to 1000 ml in a volumetric flask
4. 0.4M H_2SO_4 solution: Take 39.2 ml of H_2SO_4 in a beaker and add it drop wise in distilled water to make volume 1000 ml.
5. 66.7 mM Potassium dichromate solution: Dissolve 19.6 g of Potassium dichromate in distilled water and make volume to 1000 ml.
6. 25 mM 1, 10 Phenanthroline ferrous sulphate : Dissolve 4.95 g of Phenanthroline ferrous sulphate in distilled water and make volume to 1000 ml.
7. H_2SO_4: H_3PO_4 2:1
8. Mercuric oxide
9. 350 ml plastic bottles

Procedure

Fumigation

1. Wet the soil and incubate for 10 days at 37° C.
2. Take 50 g of oven-dried soil in a 50 ml beaker and extract immediately with 200 ml of 0.5M Potassium sulphate. In the other set, fumigate it in a wet filter paper moistened dessicator containing 25 ml of Ethanol free chloroform in a small beaker.
3. Evacuate the dessicator until chloroform has boiled for 2 minutes and place in dark at 25° C.
4. After 24 hours, remove the residual chloroform vapours in soil by repeated evacuation before extraction.

Extraction

1. Transfer soil to 350 ml plastic bottle.
2. Add 200 ml of Potassium sulphate and is keep for shaking for half an hour.

3. Filter it through the Whatman Filter Paper 42. Take 8 ml of filtrate.
4. Add 2 ml of 66.7 mM Potassium dichromate, HgO (70 mgs) and a mixture of 15 ml H_2SO_4 and H_3PO_4 (2:1).
5. Boil the mixture carefully under reflux for 30 minutes. Allow cooling and diluting with 20-25 ml distilled water.
6. To determine excess dichromate add 0.3 ml of 25 mM 1,10 phenanthroline ferrous sulphate complex solution (indicator) and titrate with ferrous ammonium sulphate solution until the colour changes from yellowish green to green.
7. Treat controls and reagent blank in the same way.
8. Standardize acidified ferrous ammonium sulphate solution against 66.7 mM Potassium dichromate.

After titration extractable C (EC) is calculated assuming that

1 ml of 66.7 mM $K_2Cr_2O_7$ = 1200 µgC

Soil microbial biomass carbon (BC) is calculated as

BC =2.64 x EC

Where,

EC is organic C extracted from fumigated soil, minus organic C extracted from unfumigated soil.

References

1. Jenkinson D.S. and Powlson D.S. (1976).The effects of biological treatments on metabolism in soil. V. A method for measuring soil biomass. *Soil Biol. Biochem.* **8**: 209-13.
2. Voroney R.P.,Winter J.P. and Beyart R.P. (1993) Soil microbial biomass C and N. In M.R. Carter (ed.) Soil Sampling and Methods of Analysis, Lewis, Chelsea, pp 277-86.
3. Vance E.D., Brookes P.C. and Jenkinson D.S. (1987) An extraction method for measuring soil microbial biomass C. **19** (6): 703-07.

Questions

1. Why Phenanthroline ferrous sulphate is used in microbial biomass analysis?
2. What is the function of Potassium dichromate and Mercuric oxide?

28

Determination of Labile Carbon in Soil

Introduction

The balance between soil organic matter (SOM) levels and nutrient cycling are the key to sustained productivity of agricultural system. Both are closely related through microbially driven mobilisation/immobilisation processes. Three major models have been recognised for C pool viz. Labile C, non-labile C and recalcitrant. The models have been characterised on the bases of their half-life period. The labile C includes oxidisable C (with KMnO4), microbial biomass C (soil carbohydrates) and soil fauna exudates (polysaccharides). Its half-life period is from a few hours to a few days. The non-labile C includes aggregate C and its turn over time is from a few months to a few years while the recalcitrant (stabilised C) includes humic fractions and charcoal C having the turn over time of hundred years. Thus fractionation of SOM is carried out on the bases of humic substances, dissolved organic C, natural C13 abundance, microbial biomass C and ease of oxidation of C. The latter method provides qualitative characterisation of soil C.

The measurement of labile C involves the ease of oxidation of the soil organic C by Potassium Permanganate. The changes in the total C in soil and its lability forms the bases of development of carbon management index.

Materials

1. 33 mM KMnO4: Weigh 5.215 g KMnO4 and place in a 1 litre Erlenmeyer flask. Add 800 ml of deionised-distilled (DD) water and a magnetic bar and place on a hot plate stirrer. Apply low heat if needed to completely dissolve the crystals. Cover the entire flask with aluminium foil to exclude light. Cool the solution slowly and transfer to a 1 litre volumetric flask and make up to 1 litre using the washings from the Erlenmeyer flask. Transfer to an amber reagent bottle.
2. 20% NaOH: Add 20 g of NaOH pellets to 60 ml of DD water. Dissolve it and then transfer to a 100 ml volumetric flask and dilute to volume.
3. 0.00025M KI: Weigh 0.0415 g KI and make to 1 litre with DD water.

4. As_2O_3
5. Concentrated HCl
6. Spectrophotometer
7. Horizontal shaker or an end-over-end tumbler
8. Centrifuge
9. 50 ml Digital burette
10. Digit rate timer
11. Volume dispenser- Pipetman kit

Procedure

Determination of the exact concentration of KMnO4

1. Weight 0.05g of As2O3 and transfer it to a 125 ml Erlenmeyer flask. Prepare 3 replicates.
2. Dissolve in 2 ml of 20% NaOH
3. Add 20 ml of double distilled water.
4. Add 2 ml of conc. HCl.
5. Add 4 drops of 0.00025M KI with a magnetic bar.
6. Stir using a magnetic stirrer and add 33mM KMnO4 from a burette while stirring until the end point is reached (faint pink colour which persists for 30 seconds)
7. Calculate the exact concentration of KMnO4 using the Equation:
   mM KMnO4 = (gAs2O3 x 1000)/(mlKMnO4x0.247275)

Preparation of standards

1. Transfer 0.6, 1.0, 1.4, 1.8 and 2.0 ml aliquots of the 33mM KMnO4 using a "Pipetman" and dilute to 50 ml using a volumetric flask.
2. The concentration of the diluted KMnO4 is computed as follows: ml aliquot x exact concentration of KMnO4 (from titration)/50 ml
3. Measure the absorbance of diluted KMnO4 standards at 565 nm. Use DD water to adjust the spectrophotometer to zero.
4. Plot absorbance (x) vs. concentration of diluted KMnO4 (approx. 0.4-1.4 mM) (y) and determine the regression equation. Range of absorbance should be between 0.5-1.7.

Determination of labile carbon in soil samples

1. Weigh air-dried soil containing 15 mg C into 50 ml centrifuge tubes with caps.

2. Add 25 ml of 33mM KMnO4 using a volumetric pipette to each soil sample.
3. Pipette 25 ml of 33mM KMnO4 into an empty centrifuge tube and use this as the reagent blank (prepare 3 reagent blanks)
4. Put the caps on the centrifuge tubes containing the soil samples and blanks. Now completely cover them with aluminium foil to exclude light. Shake the tubes using a horizontal shaker.
5. Centrifuge at 2000 rpm for 5 minutes.
6. Carefully pipette out 2 ml aliquot from each sample and blank using a "Pipetman" into 50 ml volumetric flasks and dilute to volume.
7. Set the spectrophotometer at 565 nm. Adjust to zero absorbance using double distilled water. Measure the absorbance of diluted reagent blanks and KMnO4 from soil samples.
8. Use the standard regression curve to determine the concentration of KMnO4 from the samples and reagent blanks.
9. Compute labile C as follows:

$$\text{LC (mg/g)} = \frac{(\text{mM Blank} - \text{mM Sample})\text{x } (50/2) \text{ x}25\text{x}9}{1000 \text{ x weight of soil (g)}}$$

mM Blank and mM Sample is determined from the Standard Regression Curve, 50/2 is the dilution factor, 25 is the volume of KMnO4 added to the soil sample, 9 is the factor used to convert mM KMnO4 to mg C.

References

1. Blair G.J., Lefroy R.D.B. and Lisle L. (1995) Soil carbon fractions based on their degree of oxidation, and the development of a carbon management index for agricultural systems. Australian J. Agril. Res. 46 (7): 1459-66.
2. Loginow W., Wisniewski W., Gonet S.S. and Ciescinska B. (1987) Fractionation of organic carbon based on susceptibility to oxidation. Polish J. Soil Science. 20: 47-52.

Questions

1. What is the function of As2O3 in determination of labile C?
2. Why centrifuge tubes having KMnO4 should be covered with aluminium foil?

29

Carbon and Nitrogen in Organic Matter

Introduction

Organic matter in sediment consists of carbon and nutrients in the form of carbohydrates, proteins, fats and nucleic acids. Bacteria quickly eat the less resistant molecules, such as nucleic acids and many of the proteins. Sediment organic matter is derived from plant/animal detritus, bacteria or plankton formed *in situ*. Total organic carbon refers to the amount of organic matter preserved within sediment.

Materials

1. Sample
2. Crucibles
3. Furnace

Procedure

Take 1.0 g of sample in a crucible whose empty weight is already recorded. Place this crucible in a furnace. Switch on the furnace and set the temperature at 550°C. Keep the samples for 3 hours and then switch off the furnace. Draw the samples from the furnace on next day and weigh it.

Weight of empty crucible = a

Weight of sample = b

Weight of crucible after ignition= c

Amount of ash=(c-a) x100/b

Amount of organic matter=100-(c-a) x100/b =Y

Amount of organic Carbon=Y x 0.56%

Determination of total Nitrogen

Principle

In the presence of sulphuric acid, potassium sulphate and cupric sulphate, amino nitrogen of many organic materials is converted to ammonium sulphate.

After addition of base, the ammonia is distilled from an alkaline medium and absorbed in boric acid forming ammonium borate making weak boric acid alkaline. The ammonium borate is back titrated with standard sulphuric acid until the reaction mixture changes its colour from green to pink due to formation of ammonium sulphate. The standard mineral acid used in back titration of ammonium borate is a measure of nitrogen content of the plant sample.

Materials

1. Sample
2. Digestion tube
3. Distillation unit
4. Top burette
5. Flasks
6. Catalyst mixture ($CuSO_4$: K_2SO_4: Selenium powder (20:3:1) + 500 g Na_2CO_3)
7. 4% Boric acid in distilled water (add 20 ml of mixed indicator to 1 litre of 4% Boric acid till the pH reaches 4.5-5.1))
8. NaOH (40%)
9. N/35 H_2SO_4

Procedure

1. Digestion tube containing 0.25 g of dried ground shoot material, 5 ml concentrated H_2SO_4 and 1.2 g of catalyst mixture is digested till clear.
2. The digested contents are cooled and added approximately 5 ml of distilled water. The whole sample is distilled in Micro kjeldahl distillation apparatus by adding 10 ml of 40% NaOH.
3. The liberated ammonia is trapped in 5 ml boric acid with mixed indicator.
4. About 30 ml of green distillate collected in 50 ml conical flask is titrated with N/35 H_2SO_4 till the distillate changed from green to pink.
5. Blank (without sample) is also run simultaneously.

Calculations

The N content is calculated by using the following equation

%N in sample = (a-b) x 0.16

(a-b) = Final-initial burette reading of N/35 H_2SO_4

Reference

Loginow W., Wisniewski W., Gonet S.S. and Ciescinska B. (1987) Fractionation of organic carbon based on susceptibility to oxidation. *Polish J. Soil Science.* **20**: 47-52.

Questions

1. What is the function of catalyst mixture in the estimation of Nitrogen in organic matter?
2. Why there is a need to digest the shoot material for Nitrogen content determination?
3. How do you calculate factor 0.16 in determination of %N of organic matter?

30

Estimation of Biochemical Constituents Like Cellulose, Hemicellulose, Lignin and Ash from Organic Manures/Plant Materials

Introduction

Sugars are commonly found in the form of biopolymers. Cellulose, a biopolymer of glucose is the predominant constituent of plant tissues, and may be shortly described as the structural bases of vegetable world. In this case, the glucose moieties are strung together by β-glycosidase linkages, which form the linear chain, making it resistant to chemical attack. Hemicellulose consists of short, highly branched, chains of sugars. It contains five carbon sugars (usually D-Xylose and L-Arabinose) and six carbon sugars (D-Galactose, D-Glucose and D-Mannose) and uronic acid. Its branched nature renders hemicellulose amorphous and relatively easy to hydrolyse to its constituent sugars. Lignin is a biopolymer rich in phenolic components, which provide structural integrity to plants. Acid and neutral detergent fraction methods (ADF and NDF) involve the gravimetric determination of residue previously treated with acid and neutral detergent solutions. These methods determine the insoluble fraction and their individual components. NDF represents the total plant fibre or cell wall including hemicellulose, cellulose and lignin. Acid detergent solution can be used to measure acid detergent fibre, which contains cellulose and lignin.

Principle

Neutral detergent fraction (NDF) in plant sample after heating, refluxing, filtering, washing with acetone and drying at 100^{0}C for 8 hours result in dissolving soluble constituents viz. sugars, amino acids, vitamins etc while the acid detergent fraction (ADF) in sample after passing through all these steps help to dissolve hemicellulose along with all other soluble components viz. sugars, amino acids, vitamins etc. Thus, the NDF contains hemicellulose, cellulose, lignin and ash while the ADF contains cellulose, lignin and ash. Covering ADF with 72% H_2SO_4 results in dissolving cellulose. Ignition of sample at 550^{0}C results in loss in weight which is taken as equivalent to lignin and the remaining is taken as ash content.

Materials

1. Sample
2. Flasks
3. Condensers
4. Crucibles
5. Vacuum pump
6. Oven
7. H_2SO_4 72% weight solution
8. Acetone
9. $BaCl_2$
10. NDF (Neutral Detergent fraction) solution: Solution is prepared by adding 30g/lt. of sodium lauryl sulphate, 18.6g/lt of disodium ethylene diamine acetate crystals, 6.8g/lt of sodium borate decahydrate (borax), 4.6g/lt of disodium hydrogen phosphate and 10 ml of 2-Ethoxy ethanol in 1 litre. distilled water. Adjust the pH 7.1-9.7.
11. ADF (Acid detergent fraction) solution: Add 20g of acetyl trimethyl ammonium bromide and 2 ml of decalin in 1N H_2SO_4. Make the final volume to 1 litre.
12. 72% H_2SO_4 (w/w): Specific gravity of H_2SO_4 = 1.835 g = 1.84. Thus for 1.84 g we need 1 ml H_2SO_4

    For 72 g = $\frac{1\times 72}{1.84}$ = 39.1 ml

    Add 39.1 ml of conc. H_2SO_4 in distilled water drop-wise and make the volume 100 ml. This is 72% H_2SO4 (w/w).

Procedure

Soluble constituents

Take 1.0 g of sample (rice straw) in 250 ml conical flask fitted with B 24/29 glass joint. The samples should be taken in 6 flasks separately. Add 100 ml of NDF solution to 3 flasks and 100ml of ADF solution to remaining 3 flasks. Keep the flask on the hot plate, fix the condenser and water pipe on the flask and supply the water in the condenser. Boil the contents for 5-10 minutes and then heat it gently for about 1 hour to reflux the samples. Filter the samples in already weighed glass crucibles (Sintered glass of G1, grade mounted on a suction flask). Suck the material with vacuum pump in order to facilitate the filtration. Wash the samples with hot water till foam formation is stopped in the filtrate. Wash the content with acetone (acetone helps in the easy drying of the material). Dry the crucibles at 65°C for 8 hours and then weigh it. Loss in weight in case of NDF fraction was taken equivalent to soluble constituents and left over was taken as NDF fraction containing hemicellulose, cellulose lignin and ash. The ADF fraction contains cellulose, lignin and ash.

Hemicellulose

It is equivalent to the difference in weight of NDF fraction and ADF fraction.

Cellulose

Cover the crucibles containing ADF fraction with 72% H_2SO_4 (w/w) for 3 hours and wash with water to get rid of acids. It can be judge by taking a few drops of the filtrate in a solution of $BaCl_2$. Presence of white precipitate will indicate the presence of sulphate and the sample should be washed again and again till the filtrate is clear from the sulphate content. Dry the samples in an oven for 8 hours at 65°C. Weighed the crucible and loss in weight is taken as weight of cellulose.

Lignin

Ignite the residue left after above treatment at 550°C for 3 hours in furnace. The loss in weight was taken as lignin content.

Ash

The remaining content in the crucible (Z) is taken as ash content.

Calculations

NDF

Weight of empty crucible = a1

Weight of empty crucible and sample = b1

Weight of crucible after drying = c1

Weight of sample = b1-a1 = X

Weight loss = b1-c1 = X1

NDF soluble fraction = (X1/X) x 100 = A%

ADF

Weight of empty crucible = a2

Weight of empty crucible and sample = b2

Weight of crucible after drying = c2

Weight of sample = b2-a2 = X2

Weight loss = b2-c2 = X3

ADF soluble fraction = (X3/X2) X 100 = A1%

Hemicellulose content = A1-A = P%

After acid washing and drying at 67^0 C for 8 hours

ADF = B

ADF = c2-B = C

Cellulose% = (C/X2) x 100 = Q%

Loss in weight after igniting sample (ADF) = c2-L = N

Lignin (ADF) = (N/X2) x 100 = R%

Ash = 100- (P+Q+R) = S%

References

Perez Hidalgo M., Guerra Hernandez E. and Garcia Villanova, B. (1997) Determination of insoluble dietary fibre compounds: cellulose, hemicellulose and lignin in legumes. *Ars Pharmaceutica*, **38** (4): 357-64.

Questions

1. What is the function of NDF and ADF?
2. Why sintered crucible of G1 grade is used for filtration?

31

Determination of Biological Oxygen Demand and Chemical Oxygen Demand

Introduction

In recent years, the boom in industrialization has revolutionised the countries but has also added much of sufferings. There is tremendous increase in the water and air pollution. Large amounts of effluents from industries containing organic substances of different nature are discharged and enter the aquatic systems resulting in water pollution. Biological Oxygen Demand (BOD) is the amount of dissolved oxygen required by the aerobic microflora to oxidise the organic matter of dirty water. Increased organic matter results in more oxygen requirements by microorganisms to oxidise it and it results in enhancement of BOD of water. As the BOD of water rises, due to decrease in dissolved oxygen (DO), there is increase in death of fish due to suffocation. Determination of biological oxygen demand (BOD) only is insufficient to give a clear picture of organic content of water bodies and the presence of toxic chemicals in the sample so the chemical oxygen demand (COD) is estimated. COD is a measure of capacity of water to consume oxygen during the decomposition of organic matter and the oxidation of inorganic chemicals such as ammonia and nitrate. The values of BOD and COD cannot be directly compared with each other.

Biological Oxygen Demand (BOD)

Principle

Under acidic conditions, dissolved oxygen binds with Magnesium hydroxide. It results in release of Iodine equivalent to the Oxygen fixed. The released Iodine is further titrated against sodium thiosulphate by using starch as indicator.

Materials

1. Dirty water sample
2. Clean water sample
3. BOD bottles

4. Allylthiourea solution (0.5%)
5. Manganese sulphate (50% solution)
6. Starch indicator
7. Alkali Potassium Iodide
8. Acid and Alkali solution (1N)
9. Titration apparatus
10. Pipettes (2ml)
11. Sodium thiosulphate (0.025N)

Procedure

1. Collect 250 ml samples of dirty water and clean water in BOD bottles of 300 ml capacity.
2. Adjust the pH of both the water samples to neutral by using 1N acid or alkali solution.
3. To prevent nitrification, add 1 ml of allylthiourea solution to each bottle.
4. Add 2 ml each of Manganese sulphate and alkali Potassium Iodide solutions to BOD bottles and cork them.
5. A brown coloured precipitate is developed at the bottom of BOD bottles. Add 2 drops of starch indicator solution to each. To dissolve precipate, add 2-3 ml of 1N acid solution to the bottles. Shake the bottles properly.
6. Take 2 clean flasks and add add 50 ml of sample from each of two bottles. Titrate them using 0.025M sodium thiosulphate solution till the colour changes from blue to yellow. Note the volume of sodium thiosulphate consumed. Calculate the amount of dissolved oxygen.
7. Incubate the BOD bottles again at 27^0C for 4 days and after 4 days determine the amount of dissolved oxygen again by repeating the steps 4-6.

Observations and Results

The amount of dissolved oxygen (DO) (mg/litre) can be determined as

$$\text{DO (mg/l)} = \frac{8 \times 1000 \times N}{V} \times V$$

Where,

V = Volume of water sample used for titration

v = Volume of Sodium thiosulphate consumed

N = Normality of Sodium thiosulphate i.e., 0.025N

8 = Constant as 1 ml of 0.025N of Sodium thiosulphate is equivalent to 0.2 mg oxygen.

Now calculate the BOD of water by using the following formula:

BOD (mg/l) = D1-D2

D1 = Initial dissolved oxygen (mg/l) in first sample

D2 = Dissolved oxygen in the sample (mg/l) after 4 days of incubation.

Chemical Oxygen Demand (COD)

Principle

The Chemical Oxygen Demand (COD) determines the amount of oxygen required for oxidation of organic matter & inorganic matter using a strong chemical oxidant such as potassium dichromate under reflux conditions. The excess amount of oxygen reacts with Potassium Iodide and releases Iodine. Starch solution is used as an indicator. Iodine is titrated with sodium thiosulphate and the amount is determined.

Materials

1. Dirty water sample
2. Conical flasks (100ml)
3. Potassium dichromate solution (0.1N)
4. Sulphuric acid solution (2 M)
5. Sodium thiosulphate (0.1 M)
6. Starch solution (1%)
7. Water bath
8. Distilled water blanks (3)
9. Titration unit

Procedure

1. Collect 50 ml of dirty water sample and distilled water sample each in a 100 ml conical flask.
2. Put 5 ml of 0.1 N Potassium dichromate solution in both the flasks.
3. Keep the flasks in boiling water bath for 60 minutes.
4. Allow the flasks to cool for 10 minutes
5. Add 5 ml potassium iodide solution and 10 ml sulphuric acid solution in each flask.
6. Titrate these flasks against 0.1 M sodium thiosulphate solution till the yellow colour disappears. Note the amount sodium thiosulphate solution used.

7. Add 1ml of starch solution to the flasks. The blue colour appears.
8. Again titrate the solution with 0.1 M sodium thiosulphate till the blue colour disappears.
9. Note the volume of sodium thiosulphate solution used in both dirty water and distilled water.

Calculation

$$\text{COD (mg/l)} = \frac{8 \times \text{Cx}(V_B - V_A)}{V_S}$$

Where:

C = concentration of sodium thiosulphate (titrant)(mM/l)

V_A= volume (ml) of titrant used for distilled water blank

V_B= volume (ml) of titrant used for dirty water sample

V_S= volume (ml) of water sample taken

References

1. Kocher G.H. (2012) Practical teachings in Microbiology. Narendra Publishing House, New Delhi (India)
2. Tiwari, R.P., Hoondal, G.S. and Tewari, R. (2004) Laboratory techniques in Microbiology and Biotechnology. Abhishek Publications, Chandigarh (India)

Questions

1. How do BOD and COD differ?
2. Why water samples contained in BOD bottles should be free from bubbles?
3. How COD gives satisfactory estimate of organic matter present?
4. Why the values of COD more than BOD values?

32

Quantitative Examination of Microbes in Food Products

Introduction

Both beneficiary and deleterious microorganisms are present in food. Beneficiary microbes are used for the preparation of a number of dairy and meat products and are called Probiotics, whereas deleterious microbes can cause food poisoning. The quantitative analysis of food products is necessary to find out the presence of types of microorganisms.

Requirements

1. Food sample
2. Mechanical blender
3. Sterilised distilled water
4. Steilised distilled water blanks (9 ml)
5. Petri plates
6. Sterile Pipettes (1 ml)
7. Eiosin Methylene Blue agar medium
8. Incubator
9. Colony counter
10. Marker

Procedure

1. Weigh 1 g of well blended food sample and transfer in 9 ml sterile distilled water to obtain a 10^{-1} dilution.
2. Shake this mixture vigorously for approximately 5-10 minutes. Transfer 1 ml of suspension to 9 ml water blank, it establishes a 10^{-2} dilution. From this second dilution, similarly transfer 1 ml quantities to water blank of 9 ml to provide dilution series ranging from 10^{-3} and 10^{-4} dilutions.
3. From the highest dilution prepared (10^{-4}) transfer 1 ml portion of the freshly agitated suspension to each 3 sterile Petri dishes by means of sterile 1 ml

pipette. Make similar transfers from the next lower dilution (10^{-3}, 10^{-2}) into other dishes. Inoculate 1 ml blank in 3 petri dishes. It will serve as control.

4. Into each seeded petri dish pour about 18 ml EMB agar, which previously has been steamed sufficiently to ensure complete melting and then cooled to 42^0 C.
5. Immediately after adding agar, rotate the dish by hand to distribute the inoculum throughout the medium.
6. Permit the agar to solidify. Then invert them, and place them in an incubator at 35^0C for 24 hours.

Observations and Calculations

If the plates from highest dilution show more than 300 colonies, the dilution has been too low and if those from lowest dilution show less than 30 colonies the dilution has been too high. In either case discard all plates. Select the plates from the dilution at which 30 to 300 colonies have developed. Calculate the results by multiplying average number of colonies per plate for a given dilution by the dilution factor. If there is metallic sheet on the EMB agar, it indicates the presence of *E. coli* which demonstrates the probability of faecal contamination in food.

S.No.	Dilution	Number of colonies
1	10^{-2}	
2	10^{-3}	
3	10^{-4}	

Total viable count per milliliter of food:

Reference

Kocher G.H. (2012) Practical teachings in Microbiology. Narendra Publishing House, New Delhi (India)

Questions

1. Why is EMB agar media preferred for quantification of microbes in food?
2. Name the organisms which can cause food poisoning?

Detection Methods for Microbial Quality of Milk and Milk Products

Microbial deterioration of milk and milk products occur as a consequence of either growth of microbes or release of extracellular or intracellular enzymes, toxins into them or in the environment. The microbial deteriorative changes are exhibited by changes in colour, odour, texture, formation of slime, gases, exudates etc. in milk and milk products. The microbial growth is also important with the food safety and biosecurity issues. Milk products are prone to deteriorative changes due to typical product characteristics of high water activity, nutritive value, nearly optimum pH required for growth of the spoilage microorganisms. In addition, the emergence of newer pathogens like *Listeria monocytogenes*, *Campylobacter jejuni*, *Salmonella typhimurium* etc. lead to increase in milk/food-borne illness. These pathogens have no boundary limitation and in turn circumnavigate the globe posing a serious challenge to the Government and Public health authorities.

In addition to being a nutritious food for humans, milk provides a favourable environment for the growth of microorganisms. Yeasts, moulds and a broad spectrum of bacteria can grow in milk, particularly at temperatures above 16°C.

There is an increasing focus on milk quality and hygiene in the dairy industry. Producing high quality milk requires effective udder health programs at a herd level. The safety of milk is an important attribute for consumers of milk and dairy products. Milk and products derived from milk of dairy cows can harbor a variety of microorganisms and can be important sources of foodborne pathogens. Microbes can enter milk via the cow itself, air, feedstuffs, milk handling equipment and the milker. Once microorganisms get into the milk their numbers increase rapidly. It is more effective to exclude micro-organisms than to try to control microbial growth once they have entered the milk. Milking equipment should be washed thoroughly before and after use—rinsing is not enough. Entry of foodborne pathogens via contaminated raw milk into dairy food processing plants can lead to persistence of these pathogens in biofilms, and subsequent contamination of processed milk products and exposure of consumers to pathogenic bacteria. Inadequate or faulty pasteurization will not destroy all foodborne pathogens. Furthermore, pathogens can survive and thrive in post-pasteurization processing environments, thus leading to recontamination of dairy products. These pathways pose a risk to the consumer from direct exposure to foodborne pathogens present in unpasteurized dairy products.

The safety of dairy products with respect to food-borne diseases is of great concern around the world. This is especially true in developing countries where production of milk and various milk products takes place under unsanitary conditions and poor production practices. A major factor determining milk quality is its microbial load. It indicates the hygiene practiced during milking, like cleanliness of the milking utensils, condition of storage, manner of transport as well as the cleanliness of the udder of the individual animal. Milk from a healthy udder contains few bacteria but it picks up many bacteria from the time it leaves the teat of the cow until it is used for further processing. These microorganisms are indicators of both the manner of handling milk from milking till consumption and the quality of the milk. Raw milk produced under hygienic conditions from healthy animals should not contain more than 30,000 microbes/ml. The detection of coliform bacteria and pathogens in milk indicates a possible contamination of bacteria either from the udder, milk utensils or water supply used. Fresh milk drawn from a healthy cow normally contains a low microbial load (less than 1000 ml^{-1}), but the loads may increase up to 100-fold or more once it is stored for sometimes at normal temperatures. However, keeping milk in clean containers at refrigerated temperatures immediately after milking process may delay the increase of initial microbial load and prevent the multiplication of microorganisms in milk between milking at the farm and transportation to the processing plant. Contamination of mastitis milk with fresh clean milk may be one of the reasons for the high microbial load of bulk milk. Therefore, the assessment of microbial quality of milk should be carried out considering all the independent factors that are able to affect the food safety level of the end product of the whole dairy chain, i.e., the consumed fluid pasteurized milk, transportation between the stages, i.e., transport of raw milk to the processing factory, and delivery of pasteurized milk to the sale unit (retailer/catering establishment).

For collection of the samples, the udder was washed with antiseptic solution, wiped dry with clean cloth and then disinfected with cotton ball dampened with 70% alcohol, the foremilk was discarded and 20 ml of pooled milk was collected (5 ml from each quarter).

The qualitative and quantitative estimation of microbiological quality of food products including milk and milk products depends upon the systemic collection of samples using scientific protocols.

Sampling Plans: A sampling plan is a statement of the criteria of acceptance applied to a lot based on appropriate examinations of a required number of sample units by specified methods. It consists of a sampling procedure and decision criteria and may be a two-class or a three-class plan.

A two-class plan consists of the following specifications: **n**, **c**, **m**; a three-class plan requires **n**, **c**, **m**, and **M**, where n = the number of sample units (packages, beef patties, and so forth) from a lot that must be examined to satisfy a given sampling plan. c = the maximum acceptable number, or the maximum allowable number of sample units that may exceed the microbiological criterion m. When this number is exceeded, the lot is rejected. m = the maximum number or level of relevant bacteria per gram; values above this level are either marginally acceptable or unacceptable. It is used to separate acceptable from unacceptable foods in a two-class plan, or, in a three-class plan, to separate good quality from marginally acceptable quality foods. The level of the organism in question that is acceptable and attainable in the food product is m. In the presence/absence situations for two-class plans, it is common to assign m = 0.

For three-class plans, m is usually some nonzero value. **M** = a quantity that is used to separate marginally acceptable quality from unacceptable quality foods. It is used only in three-class plans. Values at or above M in any sample are unacceptable relative to health hazard, sanitary indicators, or spoilage potential. A two-class plan is the simpler of the two and in its simplest form may be used to accept or reject a larger batch (lot) of food in a presence/absence decision by a plan such as n=5, c= 0, where n=5 means that five individual units of the lot will be examined microbiologically for, say, the presence of salmonellae, and c = 0 means that all five units must be free of the organisms by the method of examination in order for the lot to be acceptable. If any unit is positive for salmonellae, the entire lot is rejected. If it is desired that two of the five samples may contain coliforms, in a presence/absence test, for example, the sampling plan would be n = 5, c = 2. By this plan, if three or more of the five unit samples contained coliforms, the entire lot would be rejected. Although the presence/absence situations generally obtained for salmonellae, an allowable upper limit for indicator organisms such as coliforms is more common. If it is desired to allow up to 100 coliforms/g in two of the five units, the sampling plan would be n = 5, c = 2, m = 10^2. After the five units have been examined for coliforms, the lot is acceptable if no more than two of the five contain as many as 10^2 coliforms/g but is rejected if three or more of the five contain 10^2 coliforms/g. This particular sampling plan may be made more stringent by increasing n (e.g., n = 10, c = 2, m = 10^2) or by reducing c (e.g., n = 5, c = 1, m = 10^2). On the other hand, it can be made more lenient for a given size n by increasing c. Whereas a two-class plan may be used to designate acceptable/unacceptable foods, a three-class plan is required to designate acceptable/marginally acceptable/unacceptable foods. To illustrate a typical three-class plan, assume that for a given food product, the standard plate count (SPC) shall not exceed 10^6/g (M) or be higher than 10^5/g from three or more of five units examined. The specifications are thus n = 5, c = 2, m = 10^5, M = 10^6. If any of the five units exceeds 10^6/g, the entire lot is rejected (unacceptable). If not more than c sample units give results above m, the lot is acceptable.

Unlike two-class plans, the three-class plan distinguishes values between m and M (marginally acceptable). With either two- or three-class attribute plans, the numbers n and c may be employed to find the probability of acceptance (Pa) of foods. The decision to employ a two-class or three-class plan may be determined by whether presence/absence tests are desirable, in which case a two-class plan is required, or whether count or concentration tests are desired, in which case a three-class plan is preferred. The latter offers the advantages of being less affected by nonrandom variations between sample units and of being able to measure the frequency of values in the m-to-M range.

Prior to the development of the HACCP and sampling plan concepts, microbiological criteria (generally referred to as standards at the time) were applied to a variety of products.

Microbiological criteria for Dairy Products include Grade A a) raw milk for pasteurization: Not to exceed 100,000 bacteria per milliliter prior to commingling with other producer milk; and not exceeding 300,000 per milliliter milk prior to pasteurization. b). Grade A pasteurized milk and milk products (except cultured products): Not over 20,000 bacteria per milliliter, and not over 10 coliforms per milliliter. c). Grade A pasteurized cultured products: Not over 10 coliforms per milliliter.

Note: Enforcement procedures for (a), (b), and (c) require a three-out-of-five compliance by samples.

Whenever two of four successive samples do not meet the standard, a fifth sample is tested; and if this exceeds any standard, the permit from the health authority may be suspended. It may be reinstated after compliance by four successive samples has been demonstrated. B. Certified milk (American Association of Medical Milk Commissions, Inc.)

a. **Certified milk (raw):** Bacterial plate count not exceeding 10,000 colonies per milliliter; coliform colony count not exceeding 10 per milliliter.
b. **Certified milk (pasteurized):** Bacterial plate count not exceeding 10,000 colonies per milliliter before pasteurization and 500 per milliliter in route samples. Milk not exceeding 10 coliforms per milliliter before pasteurization and 1 coliform per milliliter in route samples.
c. **Milk for manufacturing and processing** i). Class 1: Direct microscopic clump count (DMC) not over 200,000 per milliliter. ii). Class 2: DMC not over 3 million per milliliter. iii). Milk for Grade A dry milk products: must comply with requirements for Grade A raw milk for pasteurization.
d. **Dry milk** i). Grade A dry milk products: at no time a standard plate count over 30,000 per gram, or coliform count over 90 per gram (U.S. Public Health Service). ii) Standards of Agricultural Marketing Service (USDA):

(1) Instant nonfat: U.S. Extra Grade, a standard plate count not over 35,000 per gram, and coliform count not over 90 per gram. (2) Nonfat (roller or spray): U.S. Extra Grade, a standard plate count not over 50,000 per gram; U.S. Standard Grade, not over 100,000 per gram (3) Nonfat (roller or spray): Direct microscopic clump count not over 200 million per gram; and must meet the requirements of U.S. Standard Guide. U.S. Extra Grade, such as used for school lunches, has an upper limit of 75 million per gram. iii). Dried milk: Mesophilic count: n = 5, c = 2, m = 5×10^4, M = 2×10^5 Coliforms: n = 5, c = 1, m = 10, M = 100 Salmonella: n = 15, c = 0, m = 0.

e. **Frozen desserts** have bacterial standards usually specify a maximal count of 50,000 to 100,000 per milliliter or gram. The U.S. Public Health Ordinance and Code sets the limit at 50,000 and recommends bacteriological standards for cream and milk used as ingredients.

Measures of Milk Quality

Somatic Cell Count: The number of somatic cells in milk, referred to as the somatic cell count or SCC, is used throughout the world as an indicator of milk quality. The current regulatory limit for somatic cells in milk in the United States defined in the Grade "A" Pasteurized Milk Ordinance is 7,50,000 cells per milliliter (ml). For many good reasons, there is continuing pressure from animal health advocacy groups to reduce the regulatory limit for somatic cells in milk from the current 750,000 cells per ml to 400,000 or less.

Poor quality milk has a high number of somatic cells and is an inferior product with reduced processing properties resulting in dairy products with a reduced shelf-life. On the other hand, high quality milk has a very low number of somatic cells, a longer shelf-life, tastes better and is more nutritious. One characteristic feature of cows with mastitis is a significant elevation in the number of somatic cells in milk. Milk from uninfected mammary glands contains less than 100,000 somatic cells per ml. A milk SCC of more than 200,000 per ml suggests that an inflammatory response has been elicited, a mammary quarter is infected or is recovering from an infection, and that milk has reduced manufacturing properties. It is not uncommon for milk from cows with subclinical and/or clinical mastitis to contain several hundred thousand and even millions of somatic cells per milliliter of milk. Thus, an increase in the SCC of milk is a good indicator of mastitis or inflammation in the udder. Infection of the udder by mastitis pathogens alters milk composition and reduces milk yield. Most studies that evaluated the influence of mastitis on the composition of milk used SCC as the basis for determining the infection status of udders and for determining the degree of inflammation.

The bulk tank SCC (BTSCC) can be used to gauge the udder infection status of a dairy herd and also gives a good indication of the loss in milk production in a herd due to mastitis. As the BTSCC increases, the percentage of mammary quarters infected increases and the percentage production loss increases. Small increases in SCC can impact production. Most herd milk contains between 200,000 to 500,000 somatic cells per ml of milk. These herds are losing at least 8% in potential milk production. Thus, methods of mastitis control that reduce SCC will not only improve milk yield and composition but will also decrease economic losses due to mastitis.

Standard Plate Count (SPC): The SPC is an estimate of the total number of viable aerobic bacteria present in raw milk. This test is done by plating milk on a solid agar, incubating plates for 48 hours at 32°C (90°F), followed by counting bacteria that grow on plates. The SPC is used to monitor progress since consistent application of proper milking system, cleaning practices, proper milking practices, udder hygiene and good mastitis prevention and control practices should allow dairy producers to produce milk with a low SPC, which is less than 5,000 colony forming units (cfu)of bacteria per milliliter. USDA regulations defined in the Pasteurized Milk Ordinance mandate that the milk SPC should not exceed 10,000 cfu/mL. Though it is impossible to eliminate all sources of bacterial contamination of milk. milk from clean, healthy cows that has been properly collected generally has a SPC less than 1,000 cfu/ml. Consistent application of proper milking practices, udder hygiene and good mastitis prevention and control practices should allow dairy producers to produce milk with a SPC of ≤5,000 cfu/mL, while most farms can produce milk with counts of less than 10,000 cfu/mL. High bacteria counts (more than 10,000 cfu/ml) suggest that bacteria are entering milk from a variety of possible sources. The most frequent cause of high SPCs is poor cleaning of milking systems. Milk residues on equipment surfaces provide nutrients for growth and multiplication of bacteria that contaminate milk of subsequent milking. Cows with mastitis (streptococcal and coliforms), soiled cows, unsanitary milking practices, failure to cool milk rapidly to <4.4°C (40°F), failure of the water heater, and extremely wet and humid weather can also contribute to high SPCs in raw milk. Some limitations of the SPC method include: 1) no indication of the bacterial types present, 2) no indication of the specific source of high counts, and 3) the SPC does not give a complete count of all bacteria as some bacteria only grow at lower temperatures .

Preliminary Incubation Count (PI count): The PI count is an estimate of the number of pyschrotrophic, or cold-loving, bacteria in milk. The PI count is not a regulatory test, and results of this test are interpreted as a general reflection of milk production practices on the farm and used as a tool to identify inadequate on-farm sanitation practices and holding temperature of milk in the bulk tank. The PI count is conducted by holding milk at 55°F for 18 hours. Bacteria that grow under refrigerated conditions are enumerated using the SPC method described above. PI counts are generally higher than SPCs. Selection of a PI count cut-off and interpretations of PI count results are difficult because variability in PI counts negatively influences repeatability. Some milk plants use a specific cut-off number while others use PI counts in relation to SPC. PI counts less than 10,000 cfu/mL are considered low, while PI counts more than 20,000 cfu/mL are considered high. A PI count three to four times higher than the SPC is suggestive of potential problems related to cleaning and sanitation of the milking system or poor udder preparation before milking (Figure 1). Failure to cool milk rapidly, marginal cooling, prolonged storage times, milking cows with wet teats, and/or extremely wet and humid weather conditions may also result in high PI counts. A PI count equal or slightly higher than a high SPC (more than 50,000 cfu /mL) suggests that the high SPC is possibly due to mastitis (Figure 1). The PI count has been used by some as an indicator of the shelf-life of processed dairy products. However, research conducted at Cornell University and Penn State University has shown that the PI count alone cannot be directly correlated with the flavor quality of raw milk or quality or shelf-life of processed dairy products. PI counts are most useful with data from other tests and additional information such as farm observations and inspections.

Laboratory Pasteurization Count (LPC): The LPC, also known as the thermoduric count, is an estimate of the number of bacteria that can survive laboratory pasteurization at 62.8°C (143°F) for 30 minutes. This process destroys most of the mastitis causing pathogens, selecting for those bacteria that can survive pasteurization temperatures (thermoduric bacteria). This is not a regulatory test required by state or federal agencies; however, some milk processors perform this test to ensure quality of the final product. Bacteria not killed by pasteurization are enumerated using the SPC method. LPCs are generally much lower than. An LPC of more than 200 cfu/ml is considered high (Table 1). A high LPC is most often seen with persistent cleaning problems; faulty milking machine or worn out parts such as leaky pumps, old pipe line gaskets, inflations and other rubber parts; and milkstone deposits. Significant contamination from soiled cows can also contribute to high LPCs.

Coliform Count (CC): The CC is a test that estimates the number of bacteria that originate from manure or a contaminated environment. Milk samples are plated on Violet Red Bile agar or MacConkey's agar and incubated for 48 hours at 32°C (90°F), after which typical coliform colonies are counted. Coliform counts reflect hygiene and sanitation practices followed on the farm.

Coliforms enter the milk supply as a consequence of milking dirty cows or dropping the milking claw into manure during milking. Coliform counts >100 cfu/ml suggest poor milking practices, dirty equipment, contaminated water, dirty milking facilities, and/or cows with subclinical or clinical coliform mastitis.

Table 1 : Interpretive Criteria for Bulk Tank Milk Monitoring (EU & USDA)

Parameter	Low	Medium	High
Bulk tank SCC	<200,000	200,000-400,000	>400,000
Standard Plate Count (SPC)	<5,000	5,000-10,000	>10,000
Preliminary Incubation Count (PIC)	<10,000	10,000-20,000	>20,000
Lab Paseurized Count (LPC)	<100	100-200	>200
Coliform Count	<50	50-100	>100

Methods for the examination of milk and milk products

The examination of milk and milk products is essential for the prediction of storage stability as well as food safety point of view. The counting of number of microorganisms in the milk product is varied due to various factors such as sampling methods, distribution of organisms in the food sample, nature of food and its flora, extrinsic factors of plate; media, pH, a_w, type of diluents used, relative number of organisms in sample, existence of competitive microorganisms. The examination of microbiological quality of milk products can be carried out traditionally by **counting of microcolonies.**

Microscopic colony count methods involve the counting of microcolonies that develop in agar layered over microscope slides. In this, spread 0.1 ml of milk-agar mixture over 4 cm^2 area on glass slide. After incubation at 20-22°C, drying and staining, microcolonies are counted with the aid of a microscope.

In another method, 2 ml of melted agar are mixed with 2ml of warmed milk and after proper mixing, 0.1 ml of the inoculated agar is spread over 4 cm^2 area. Thereafter, the staining was carried out with thionine blue, the slide is viewed with 16 mm objective of a wide field microscope.

Agar Droplet method: The agar drop let method can be used to determine the number of viable suspended bacteria in a known volume. This method has some advantages over the spread plate (SP) method. Less time and effort are required to dispense the drops onto an agar plate than to spread an equivalent total sample volume into the agar. By distributing the sample in drops, colony counting can be done faster and perhaps more accurately. In this method, the sample homogenate is diluted in melted agar tubes (45°C). This test is generally performed in triplicate. For each sample take three tubes of agar, the first tube is inoculated with 1 ml of sample homogenate. After mixing, sterile capillary pipette (ideally delivering 0.033 ml/drop) is used to transfer a line of 5×0.1 ml droplets to the bottom of an empty petri dish. With the same capillary pipette, three drops (0.1 ml) from the first 9 ml test tube are transferred to the second tube and after mixing, another line of 5×0.1 ml droplet is placed next to the first. This step is repeated for the third tube of agar. Petri plates containing the agar droplets are incubated for 24 h and colonies are enumerated with the aid of 10X viewer. Results using this method from pure cultures, milk products are compared with conventional method. It is generally higher than SPC. This method is about 3 times faster and 24 h incubation is equivalent to 48 h incubation of traditional methods.

Dry films: Though, the methods for the enumeration of the mesophilic aerobic bacteria have commonly been used a traditional standard plate count method. However, this method has several disadvantages, such as time-consuming, labor-intensive, bulky waste, and cumbersome. In order to eliminate these shortcomings, dry medium culture plate methods, such as Petri-film and Sanita-Kun dry medium culture plates, could be alternatives to the traditional culture method. A dry film consisting of two plastic films attached together on one side and coated with plate count agar ingredients and cold-water soluble jelling agent has been developed. The cover film contains adhesive, a cold water-soluble gelling agent, and 2,3,5-triphenyltetrazolium chloride (TTC). Here, TTC is reduced by bacterial metabolism to produce a visible red dye. The bottom film contains a cold water-soluble gelling agent and dried nutrients. In comparison, Sanita-Kun dry medium culture plate consists of a transparent cover film, an adhesive sheet, and a layer of non- woven fabric that includes water-soluble polymers containing nutrients for microorganisms as well as TTC. In this 1 ml of diluents is placed between the two films and spread over the nutrient area by pressing. After incubation, microcolonies appear red because of the presence of TTC. On comparison of this method with traditional SPC and standard plate loop method, the correlation coefficient between dry film and SPC was 0.946 and the mean log difference was -0.177. This method is recommended as a suitable alternative to SPC and very frequently being used in the dairy industry for counting of coliform bacteria in milk and milk products.

Most Probable Number (MPN) method: This method is based upon statistical probability. Milk product samples are prepared similarly as in case of SPC. Three serial dilutions are prepared and then transferred to 9 or 15 tubes (3- or 5-tube method). The numbers of organisms/g are then estimated using standard MPN tables. This method has some innate advantages such as easy to perform, results from one lab more likely than SPC to agree with those from another lab, specific groups of organisms (e.g., coliforms-it is the method of choice) can be enumerated using different selective medias and included in AOAC for coliforms, *S. aureus* & *B. cereus* (many others) in various foods. The basic disadvantages include lots of tubes required (clean up intensive) and lack of precision, generally higher than SPC results.

The MPN method consists of taking replicate samples of product (by volume or weight) which contain, on average, the same number of viable microorganisms in each of them (hence the requirement for randomness of distribution) and scoring each sample individually for the presence of viable microorganisms by means of transferring to liquid growth media and incubation. When a sufficient volume of eluent is available, a range of dilutions can be inoculated into nutrient medium such that a fraction of the inoculated media does not produce visible growth on subsequent incubation. From the frequency of the occurrence of positive tests within a set of replicates, an estimate is made of the number of viable microorganisms present in the sample or the bulk product from which the sample has been taken; the 95% confidence limits about the estimate are relatively wide. The estimate and its confidence limits are derived from published MPN Tables, which have been developed on the assumption that numbers of viable microorganisms present in replicate samples are distributed around a mean number in accordance with Poisson distribution.

The key requirement for the application of the MPN method is the random distribution of microbial population throughout the product under investigation. Accordingly, the MPN method may have value for the determination of bioburden for liquid milk, viscous milk fluids, milk powders. However, it is difficult to determine populations of solid milk products.

Dye Reduction Test: Two dyes are commonly used in supernatants of product sample are added to standard solutions of methylene blue or resazurin:

a. Blue to White for reduction of methylene blue

b. Slate blue to pink or white for resazurin

The time required for dye reduction is inversely proportional to the total number of microorganisms. Technique lends itself to automation and has a long history of use in dairy foods. An automated colorimeter with AOAC approval has been developed for the determination of total counts and coliforms.

Advantages

1. Simple, rapid and inexpensive
2. Only viable cells reduce dye
3. Automation available

Disadvantages

1. Not all microorganisms reduce dye equally well
2. Some foods like meat contain high levels of natural reductants that cloud results. Use of a stomacher instead of a blender to homogenize meat has been shown to reduce this problem. Apparently, the stomacher doesn't release as much NADH or other reductants from the tissue.
3. Expensive equipment

   While conducting dye reduction test the samples should be mixed thoroughly and pour 10 ml. of milk sample in to previously sterilized test tube. Add quickly 1 ml. of Resazurin solution in the test tube. Mix the milk and dye thoroughly by inverting 2-3 times. Place the tubes in the water bath at the temperature of 37.5 0 C only for two minutes. The tubes are then removed from the water bath. Compare the colour to test tube with standard disc until the colours are matched under comparator. Record the number of disc/colour of disc. If colour falls between two-disc numbers record half value.

Impedance method: Impedance can be defined as the resistance to flow of an alternating current as it passes through a conducting material. Microbial metabolism usually results in an increase in both conductance and capacitance causing a decrease in impedance and a consequent increase in admittance. Impedance is a measurement based on the resistance in an electric circuit to the flow of alternating current. Microorganisms metabolize substrates of low conductivity into products of higher conductivity. As a result, resistance decreases with growing cell numbers. With proper instruments, the change in impedance can be followed and used to detect as few as 10 to 100 cells in a sample within about 6 h (impedance detection time or IDT). The technique is faster than SPC and gives results that are within 90-95% agreement with the former technique but is not AOAC approved. Measurement of impedance is widely accepted as the most versatile, the least expensive of all rapid bacterial detection methods. It is measured by an instrument which combines ease of use with leading edge electronic technology. Two specific techniques are commonly used. In the direct technique, metabolising micro-organisms increase the electrical conductance of the culture medium in the system. **Rapid Automated Bacterial Impedance Technique (RABIT)** for Windows™ measures these changes and provides results faster than by the use of traditional methods. The indirect technique provides a flexible impedance method which monitors the amount of carbon dioxide produced by growing organisms. This technique is particularly suitable for detecting organisms, which do not produce highly charged metabolites, for example yeasts and moulds. For impedance testing 1:10 dilution of milk products (cheese, paneer etc.) are prepared in 0.1% peptone water and 1.0 ml of aliquot is mixed with equal amount of appropriate growth media in sterile wells of impedance detection instrument, whereas the liquid milk and its products can be directly added into the sterile wells after thorough mixing with broth media. The samples are incubated and measured for impedance changes for 24 h. The impedance detection time (IDT) is recorded and used to estimate the viable microbial counts in the sample. The TVC is inversely related to IDT. Two methods are used calibration curve method and sterility test method. In the calibration curve method, the calibration curve is developed between IDT and SPC or TVC. Linear or quadratic regression analysis is used for analyzing the data between IDT and SPC. The sterility method was developed to detect the impedance changes in the sample due to contamination. The usefulness of this technique depends upon the large difference in the counts of acceptable and unacceptable samples. In general, the impedance of raw milk to the cut off time of 7 h (10^5cfu/ml) was used to screen the sample. To monitor the post pasteurization contamination o cream, an IDT of >20 h was found to denote cream of acceptable microbiological quality.

Microcalorimetry: Heat generation is a final by-product of all metabolic processes and therefore, its measurement gives an overall estimation of biological activity of the system. The heat is mainly the result of degradative reactions of cells, since the enthalpy change of anabolism is low or may even be negligible compared to that of catabolic processes. Any change in heat flux derived from the addition of a substrate or toxic material can be measured in order to assess biological parameters or toxicity effects. Calorimetry, as well as respirometry, can provide valuable information for modelling and control of aerobic bio-processes as heat production and respiration are linked to two important biochemical processes in wastewater treatment plants: biomass growth and substrate consumption. Moreover, because of the universality of enthalpy changes in biological reactions, calorimetric techniques can also be used in anoxic and anaerobic condition. There are two types of calorimeters- batch and flow. Microcalorimetry results vary according to the history of the organism, inoculums size, fermentable substrates, and the like. Microcalorimetry is used to study the spoilage of food products.

Flow Cytometry: Flow cytometry permits the simultaneous analysis of numerous parameters of eukaryotic cells as well as microorganisms making it a powerful tool for differentiation and functional analysis. Flow cytometry is the science f measuring components (cells) and the properties of individual cells in liquid suspension. Flow cytometers may suit the broad and specific needs of microbial analysis of milk and dairy products within one type of instrument. Flow cytometry (FCM) is extremely sensitive, avoids the need for culturing or enrichment procedures, and can be both qualitative and quantitative. Use of fluorescent stains or fluorogenic substrates in combination with FCM allows the detection and discrimination of viable culturable, viable nonculturable, and nonviable organisms. Furthermore, there is the possibility that numerous (or even rare) microbial cells could be detected against a background of other bacteria or nonbacterial particles by combining FCM and specific fluorescently-labeled antibodies or oligonucleotide probes. FCM and cell sorters make use of one or more excitation sources such as argon, krypton, or helium-neon ion lasers, and one or two fluorescent dyes to measure and characterize several thousand cells/sec. When a dye is used, its excitation spectrum must watch the light wavelengths of the excitation source. Two dyes may be used in combination to measure total protein and DNA content. For this both dyes must be excited at the same wavelength and emit at different wavelength so that the light emitted by each dye is measured separately. FCM is also used to differentiate between living and dead cells by dual staining, to determine the ploidy of yeast cells, to differentiate between spores and vegetative cells, and to separate pathogenic and non-pathogenic amoebae.

Limulus Amoebocyte Lysate (LAL) Test: Endotoxins are high-molecular-weight complexes associated with the outer membrane (lipo-polysacchride layer) of gram-negative bacteria. LAL test is used for the detection of pyrogenic endotoxins and as an alternative to the RABIT test. LAL assay is a very sensitive method for the detection of bacterial endotoxins currently approved by the FDA. LAL test employs a lysate protein obtained from the blood cells (amoebocytes) of the horse shoe crab (*Limulus polyphemous*). The lysate protein is the most sensitive substance known for endotoxins. The first methodology used to determine the LAL test results was the formation of a gel-clot in the bottom of a glass reaction tube. LAL test is a good rapid indicator of the total number of gram-negative bacteria. The method is suitable for the rapid evaluation of hygienic quality of milk relative to the detection of coliforms before and after pasteurization of milk. LAL is a method used to determine the history of milk products relative to gram-negative bacteria. Since, both viable and non-viable gram-negative bacteria are detected by LAL, a simultaneous plating s necessary to determine the number of cfu's. LAL title of liquid milk can be determined either by direct serial dilutions or by MPN with results by the two methods being essentially similar. The value of the LAL test lies in the speed at which results can be obtained. Foods that have high LAL titers can be candidates for further testing by other methods; those that have low titers may be placed immediately into categories of lower risk relative to numbers of gram-negative bacteria.

Adenosine-Triphosphate (ATP) Measurement: This test method measures the concentration of cellular-ATP present in the sample. ATP is a constituent of all living cells, including bacteria and fungi. Consequently, the presence of cellular-ATP is an indicator of total metabolically active microbial contamination in fuels. ATP is not associated with matter of non-biological origin. ATP concentration is affected by: the microbial species present, the physiological states of those species, and the total bioburden. The amount of ATP per cell is substantially greater for active fungal cells than bacteria. Within a species, cells that are more metabolically active will have more ATP per cell than dormant cells, such as fungal spores. Because fungal spores are more hydrophobic than active fungal material (mycelium), spores may be the only indicator of fungal proliferation when fuel samples are taken from some fuel systems, but they will not be detected by a test for ATP. The greater the total bioburden, the greater the ATP concentration in a sample. The ATP is measured using a bioluminescence enzyme assay, whereby light is generated in amounts proportional to the concentration of cellular-ATP in the samples. The light is produced and measured quantitatively as relative light units (RLU) which are converted by comparison with an ATP standard, computation to pg ATP/mL and optional further transformation to Log_{10}(pg ATP/mL). This test method is equally suitable for use as a laboratory or portable method. One of the simplest ways to measure ATP is by use of the firefly luciferin–luciferase system. In the presence of ATP, luciferase emits light, which is measured with a luminometer. The amount of light produced by firefly luciferase is directly proportional to the amount of ATP added. This test has an excellent potential for the rapid estimation of microorganisms in foods. The major problem that has to be overcome for food use is the removal of nonmicrobial ATP. The Triton X-100 and apyrase are selectively used to destroy nonbacterial ATP in urine specimens and subsequently the resultant ATP levels were close to values observed in laboratory cultures with detection at 10^5 bacteria per milliliter. In meats, the problem of non microbal ATP was addressed by the use of a three-stage process consisting of centrifugation, use of cation exchange resin, and filtration to get rid of food particles and collect bacteria on 0.22-μm filters. ATP analyses were carried out on bacteria eluted from the filter membranes, and 70–80% of most microorganisms were recovered on the filters. A linear relationship was shown between microbial ATP and bacterial numbers over the range 10^6–10^9 cfu/g. The ATP assay has been employed successfully for seafoods and for the determination of yeasts in beverages. The ATP assay is widely used as a rapid and on-the-spot method for monitoring food handling surfaces by swabbing designated areas and reading the relative light units (RLU) from a luminometer. Since nonmicrobial ATP can contribute to RLU readings, these methods, while valuable for monitoring purposes, should not be used to indicate numbers of microorganisms.

Fluorescent Antibody: This technique has had extensive use in both clinical and food microbiology since its development. An antibody to a given antigen is made fluorescent by coupling it to a fluorescent compound and when the antibody reacts with its antigen, the antigen–antibody complex emits fluorescence and can be detected by the use of a fluorescence microscope. The fluorescent markers used are rhodamine B, fluorescein isocyanate, and fluorescein isothiocyanate with the last being the most widely used. The fluorescent antibody (FA) technique can be carried out by use of either of two basic methods. The direct method employs antigen and specific antibody to which is coupled the fluorescent compound (antigen coated by specific antibody with fluorescent label). With the indirect method, the homologous antibody is not coupled with the fluorescent label, but instead an antibody to this antibody is prepared and coupled (antigen coated by homologous antibody, which is, in turn, coated by antibody to the homologous antibody bearing the fluorescent label). In the indirect method, the labeled compound detects the presence of the homologous antibody; in the direct method, it detects the presence of the antigen. The use of the indirect method eliminates the need to prepare FA for each organism of interest. The FA technique obviates the necessity of pure culture isolations of salmonellae if H antisera are employed. A commonly employed conjugate is polyvalent salmonellae OH globulin labeled with fluorescein isothiocyanate with somatic groups A to Z represented. Because of the cross-reactivity of salmonellae antisera with other closely related organisms (e.g., Arizona, Citrobacter, *E. coli*), false-positive results are to be expected when naturally contaminated foods are examined. The popularity of this method for foodborne pathogens has waned with the advent of molecular and other detection methods.

Immunoassays for rapid detection of pathogens: An immunoassay is a biochemical test that measures the presence or concentration of a macromolecule in a solution through the use of an antibody or immunoglobulin. The macromolecule detected by the immunoassay is often referred to as an "analyte" and is in many cases a protein. These tests rely on the specific antigen-antibody reaction and production of agglutination. This technique consists of adding a radioactive label to an antigen, allowing the labeled antigen to react with its specific antibody, and measuring the amount of antigen that combined with the antibody by the use of a counter to measure radioactivity. Solid-phase radioimmunoassay (RIA) refers to methods that employ solid materials or surfaces onto which a monolayer of antibody molecules binds electrostatically. The solid materials used include polypropylene, polystyrene, and bromo acetylcellulose. The ability of antibody-coated polymers to bind specifically with radioactive tracer antigens is essential to the basic principle of solid-phase RIA. When the free-labeled antigen is washed out, the radioactivity measurements are quantitative. Staphylococcal enterotoxin A was extracted from a variety of foods, including ham, milk products, and crabmeat and measured by RIA, all within 3–4 hours. It confirms that the method was highly sensitive and useful up to 0.001 μg/ml and quantitatively reliable up to 0.01 μg/ml of enterotoxin A. By iodination of enterotoxins, solid-phase RIA can be used to detect as little as 1 ng of toxin per gram. The RIA technique lends itself to the examination of foods for other biological hazards such as endotoxins, paralytic shellfish toxins, and the like. The detection and identification of bacterial cells within 8–10 minutes have been achieved198 by use of 125I-labeled homologous antibody filtered and washed on a Millipore membrane. Because of its requirement for an isotope and its lack of portability, the RIA method is rarely used now in the context of food microbiology.

Oligonucleotide DNA Probes: This test is based on DNA or DNA-RNA complementary base pairing or hybridization. This test can use radioactive or non-isotopic labels on probes. Reactions can require anywhere from 10 h - 2 days. This technique requires that the sequence of the target nucleic acid (frequently a gene for toxin production or a species-specific 16S rRNA sequence) be determined and unique to the organisms sought. Probes to the species-specific regions are used to detect particular bacteria such as Salmonella. One advantage to targeting the 16S rRNA versus a DNA sequence in the chromosome is that there are several copies of the 16S rRNA per cell. The same advantage applies to genes located on multicopy plasmids. Gene probes may be based on the DNA sequence of particular toxins such as *Clostridium perfringens* or staphylococcal enterotoxins.

A DNA probe consists of the DNA sequence from an organism of interest that can be used to detect homologous DNA or RNA sequences. In effect, the probe DNA must hybridize with that of the strain to be sought. Ideally, the probe contains sequences that code for a specific product. The probe DNA must be labeled in some way in order to be able to assess whether hybridization has occurred. Radioisotopes are the most widely used labels, and they include ^{32}P, ^{3}H, ^{125}I, and ^{14}C, with ^{32}P being the most widely used. Reporter groups such as alkaline phosphatase, peroxidase, fluorescein, hapten, digoxigenin, and biotin have been developed and used. When biotin is used, its presence is detected with avidin-enzyme conjugates, antibiotin, or photobiotin. Chromosomal DNA is often the source of target nucleic acid, but it contains only one copy per cell in most cases. Multiple targets are provided by mRNA, rRNA, and plasmid DNA, thus making for a more sensitive detection system. Synthetic oligonucleotide probes may be constructed of 20–50 bases, and under proper conditions, hybridization times of 30–60 minutes are possible. DNA fragments of unknown organisms are prepared by the use of restriction endonucleases. After separating fragment strands by electrophoresis, they are transferred to cellulose nitrate filters and hybridized to the labeled probe. After gentle washing to remove unreacted probe DNA, the presence of hybridized product is assessed by autoradiography. The minimum number of bacterial cells that can be detected with a standard probe is in the 10^6–10^7 range, although some investigators report the detection of 10^4 cells. When probes are used on foods where perhaps only 1 cfu/ml of the target organism exists, enrichment procedures must be employed to allow this cell to attain a level that will provide enough DNA for detection. With an initial cell number of 10^8 cells, probe results may be obtained in around 10–12 hours when radiolabels are employed. When enrichments are necessary, the time required for results would be enrichment time plus probe assay time, generally 44 hours or more. DNA probes are also used in colony hybridization methods where microcolonies of macro-colonies of the target organism are allowed to develop directly on a membrane following incubation on a suitable agar medium. A replica plate is produced as a duplicate of the master plate or membrane. Colonies that have grown on the duplicate plate are lysed directly on the membrane to release nucleic acid and to convert DNA into single strands. Some of the DNA is transferred to nitrocellulose filters, where hybridization is carried out by applying a labeled DNA or RNA probe. A modification of the traditional DNA colony hybridization technique has been made such that 60 filters with up to 48 organisms per filter can be used. The colony hybridization method has been employed successfully to detect *Listeria monocytogenes*, enterotoxigenic *E. coli*, and *Yersinia enterocolitica*. The colony hybridization was used to detect *E. coli* in artificially contaminated food without enrichment, and the method could detect 100–1,000 cells per gram, or about 1–10 cells per filter.

Polymerase Chain Reaction (PCR) based methods: PCR based methods are fast becoming popular for detecting and identifying bacteria and viruses in foods. Its increasing use is due to its high sensitivity, specificity, its availability in a number of formats, and the commercial availability of PCR-based methods in kit like formats. This technique is applicable more to the identification of foodborne organisms than to their enumeration. In this method, the general mode follows with the starting genomic material, ds DNA, it is heated to ca. 95°C to separate the strands. When the starting material is RNA (e.g. RNA viruses), it is converted to ds DNA by use of reverse transcriptase (RT-PCR). After heating to separate the DNA strands, they are cooled to 55°C in the presence of oligonucleotide primers, which anneal to the single DNA strands. DNA polymerase plus dATP, dCTP, dTTP, and dGTP are added resulting in the synthesis of complementary strands (the "d" is deoxynucleotide; A = adenine; TP = triphosphate, etc. for the other bases. These are precursors of

DNA synthesis). When this process is repeated, the two strands become four, the four become eight, and so on for each additional cycle, resulting in several million copies of the original if enough cycles are run. AG-9600 Amplisensor Analyzer is an automated fluorescence-based system for detecting PCR products. BAX is the oldest of these detection systems and its application, along with that of other PCR-based methods includes Multiplex PCR, RT-PCR, Probelia system, Molecular beacon PCR, PCR-DGGE (denaturing gradient gel electrophoresis). A number of other PCR-based methods have been developed and tested, and this is an area of active pursuit with those below being among some that have been used for foodborne organisms. LAMP—loop-mediated isothermal amplification Rep-PCR—surveys repetitive sequences only ERIC—enterobacterial repetitive intergenic consensus Molecular Beacon—uses fluorescently labeled oligonucleotides in a qPCR format QC-PCR—quantitative competitive VNTR—variable number of tandem repeats (minisatellites) Real-time PCR (qPCR, RT-PCR)—PCR results in real time. TaqMan is a fluorogenic nuclease assay that produces results in real time.

FISH (fluorescent in situ hybridization) is an RNA probe method that targets 23S rRNA. To detect *Salmonella spp*. in foods, two established oligonucleotide probes (Sal-1 and Sal-3) and a newly constructed one were used. The probes were labeled with Cy3 (indo carbocyanine) and applied to cultures of whole cells on slides.55 Fluorescence occurred when the probes hybridized with the 23S rRNA of target organisms (16S RNA sequences are more often used for bacteria). After slides were prepared, they were counter-stained with DAPI (4,6-diamidino-2-phenylindole; a non-intercalating, DNA-specific stain that fluoresces blue or bluish-white when bound to DNA and excited with light of wavelength 365 nm),109 and fluorescence was detected by epifluorescence microscopy. Although it is unclear as to the minimum number of cells needed for a positive FISH response, this number is probably $>10^4$ cfu/ml since it is a microscopic method.

LAMP (loop-mediated isothermal amplification) is an in-situ DNA amplification technique for the microscopic detection of bacteria. It is similar to FISH in that they are microscope slide methods that emit fluorescence from Cy3- or DAPI stained cells. LAMP employs a low molecular weight DNA polymerase that enters whole cells; a reaction temperature of 63^0C that makes it possible to use fluorescent antibody (FA) for simultaneous cell detection; and generates large tandem repeats that prevent amplicons from leaking out of cells. After DNA amplification using primers for the *E. coli* 0157:H7 Stx 2 gene, the identification of these cells was possible with FA labeling.

Microarrays Technique: Microarray has emerged as a promising new technology for broad-spectrum pathogen detection in clinical and environmental samples, making it possible to test for the presence of thousands of pathogens simultaneously. The advantage of microarray-based detection is that it can combine powerful nucleic acid amplification strategies with the massive screening capability of microarray technology, resulting in a high level of sensitivity, specificity, and throughput. In addition to the previously mentioned caveats, the cost and organizational complexity of performing a large number of PCR reactions for downstream microarray applications render this option feasible but unattractive. This limitation has severely reduced the utility of this technique and impeded the continued development of downstream applications.

A very simple microarray may consist of a solid surface (such as a nylon membrane, glass slide, or silicon chips) onto which are attached small quantities of single-stranded DNA (ssDNA) from different known bacterial species. When ssDNA from as many unknown species is exposed to this array (DNA chip), complementary strains will bind to their respective sites on the chip. If a reporter molecule is used, the identity of the unknown species can be confirmed. A DNA microarray is, in essence, a dot blot set-up with the capacity to obtain and process large amounts of data. A DNA microarray for microorganisms begins with the construction of oligonucleotides (primers), probes, hybridization, and data analysis. DNA array-based method gave an accurate picture of the heterogeneous bacterial community. A fiber optic DNA microarray has been developed for the detection of some foodborne pathogens, with as few as 100 cfu being detectable in <1 hour. Microarrays represent great potential for identifying microbial species and strains in foods and for the fingerprinting of biotypes.

References

1. Forrest W.W. 1972. Microcalorimetry. Methods Microbiol. 6B:385–18.

2. Frazier W.C. and Westhoff D.C. 2008 Food Microbiology. Fourth Edition, tata McGraw-Hill Publishing Company Limited, New Delhi

3. Jay J.M. 1989. The Limulus amoebocyte lysate (LAL) test. In Progress in Industrial Microbiology: Rapid Methods in Food Microbiology, ed. M.R. Adams and C.F.A. Hope, 101–19. Amsterdam: Elsevier.

4. Jay J.M., Loessner M.J., Golden D.A., 2005. Modern food microbiology. Seventh Edition, Springer Publication

5. Kepner R.L. Jr., and J.R. Pratt. 1994. Use of fluorochromes for direct enumeration of total bacteria in environmental samples: Past and present. Microbiol. Rev. 58: 603–15.

6. Khetarpaul N.,2005. Food Microbiology. Daya Publishing House, New Delhi

7. Patil S.C. and Naidu R. 2010. Food Microbiology. Campus Books International, New Delhi

8. Ray B. and Bhunia A. 2008 Fundamentals of Food Microbiology. Fourth Edition, CRC Press Taylor and Francis Group, NewYork

9. Rudi K., Flateland S.L., Hanssen J.F., Bengtsson G., and Nissen H. 2002. Development and evolution of a 16S ribosomal DNA array-based approach for describing complex microbial communities in ready-to-eat vegetable salads packed in a modified atmosphere. Appl. Environ. Microbiol. 68:1146–1156.

10. Scott T.M., Parveen S., Portier K.M., Rose J.B., Tamplin M.L., Farrah S.R., Koo A., and Lukasik J. 2003. Geographical variation in ribotype profiles of Escherichia coli isolates from humans, swine, poultry, beef, and dairy cattle in Florida. Appl. Environ. Microbiol. 69: 1089–92.

33

Bacteriological Examination of Milk By Direct Microscopic Method

Introduction

Bacteriological examination of milk indicates the hygienic condition of the milk i.e., quality of milk in relation to the health of consumer. Two main objectives of carrying out bacteriological examination are to evaluate the degree of cleanliness in the vicinity of production and handling of milk; and secondly to check the presence of infectious microorganism. Generally, number of bacteria present per milliliter is taken as an index in the routine examination of milk as other tests for identification and differentiation of pathogenic and non-pathogenic microorganisms are time consuming and expensive.

Direct Microscopic Count (DMC) consists of examining stained films of measured volume of milk or milk product, dried on a glass slide under a compound microscope. This method enables rapid estimation of total population of bacteria of a milk sample with relatively more accuracy as compared to other methods.

Requirements

1. Glass slides
2. Glass marker pen/pencil
3. Microscope
4. Stage micrometer
5. Slide holder or forceps
6. Micropipette
7. Spirit lamp
8. Needle with bent point
9. Newman's stain
   a. Methylene blue chloride - 1 g
   b. Glacial acetic acid - 6 ml

c. Tetrachloroethane - 40 ml

d. Ethyl alcohol - 54 ml

10. Cedar wood oil

Procedure

1. Clean the fresh glass slide with alcohol-soaked cotton.
2. Mark one square cm. area in the centre of glass slide.
3. Mix the milk thoroughly and transfer 0.01 ml on the marked area with the help of micropipette.
4. Spread the sample uniformly over the marked area with the help of needle/loop.
5. Dry the smear in the air.
6. Fix the stain by gently heating the slide.
7. Flood the smear with the Newman's stain.
8. Allow the stain to act for 2-3 minutes.
9. Pour off the stain and wash the slide with the clean tap water.
10. Dry the slide carefully with the help of blotting paper.
11. Put a drop of cedar wood oil on the stained smear and observe the slide under oil immersion.
12. Count the number of individual cells or clumps of bacteria in at least 25 different fields.
13. Take average of the counts and calculate the total number of bacteria/ml of milk by multiplying average counts with the microscopic factor.

Observations

<table>
<tr><td colspan="2">S.No</td><td colspan="4">Sample No</td><td colspan="19">Diameter of the microscopic field</td></tr>
<tr><td colspan="25">Counts/field</td></tr>
<tr><td>1</td><td>2</td><td>3</td><td>4</td><td>5</td><td>6</td><td>7</td><td>8</td><td>9</td><td>10</td><td>11</td><td>12</td><td>13</td><td>14</td><td>15</td><td>16</td><td>17</td><td>18</td><td>19</td><td>20</td><td>21</td><td>22</td><td>23</td><td>24</td><td>25</td></tr>
<tr><td></td><td></td><td></td><td></td><td></td><td></td><td></td><td></td><td></td><td></td><td></td><td></td><td></td><td></td><td></td><td></td><td></td><td></td><td></td><td></td><td></td><td></td><td></td><td></td><td></td></tr>
</table>

Average Count = total count/no. of fields

Calculations

1. Determination of the Microscopic factor (MF)

i. Measure the diameter of the microscope field using stage micrometer under oil immersion.

ii. Calculate the radius of the microscopic field (diameter/2)

iii. Calculate the area of the microscopic field (Πr^2, where r is the radius of the field).

iv. Determine the microscopic factor as per following formula:

$$\text{Microscopic Factor (MF)} = \frac{\text{Area of the smear}}{\text{Area of the microscopic field}} \times \frac{1}{\text{volume of milk}}$$

2. Average Number (N)

N = Total count/No. of fields counted

3. Direct Microscopic count/ml (DMC/ml)

DMC/ml. = N x M F

Direct Microscopic count/ml:	Bacterial quality of milk
< 5, 00,00	Good
5, 00,000 - 40, 00,000	Fair
40, 00, 000 – 2, 00, 00, 000	Poor
>2, 00, 00, 000	Very poor

Precautions

i. Use always clean and dry glass slides.

ii. Protect the film from dirt during drying.

iii. Dry the slide carefully.

Reference

Dubey R.C and Maheshwari D.K. (2004) Practical Microbiology. S. Chand & Company Ltd.

Questions

1. Why do microscopic factor is usually taken as 5 lakhs?
2. What are the sources of error in estimation of counts by this method?

34

Enumeration of Bacteria in Milk by Standard Plate Count Method and Presumptive Test for Coliforms

Introduction

Standard Plate Count (SPC) is used for determination of population of viable microbes. Enumeration procedures are based on the assumption that microbial cells present in a sample mixed with an agar medium form visible separated colonies. The bacterial colonies developed on the plates are counted. Each colony is presumed to have developed from one bacterium or clump of bacterial cells. Therefore, the results are expressed in colony counts per /ml of sample or colony forming units per /ml (cfu /ml). Total number of colonies counted on the plate multiplied by the dilution factor represents the number of viable organisms in a sample. SPC is a useful method for evaluating the microbiological quality at various stages of processing. It is also helpful in detecting the source of contamination in the product.

Requirements

1. Milk sample (raw/ pasteurized/ boiled)
2. Sterilised distilled water
3. Steilised blanks (9 ml)
4. Petri plates
5. Sterile Pipettes (1 ml)
6. Test tubes
7. Test tube stands
8. Conical flasks
9. Tryptone glucose yeast extract agar medium
10. Mac Conkey broth medium
11. Durham tubes
12. Incubator

13. Colony counter
14. Marker
15. Spirit lamp
16. Aluminium foil
17. Cotton plugs

Composition of Tryptone glucose yeast extract agar medium

Tryptone	5.0 g
Yeast extract	2.5g
Glucose	1.0g
Agar powder	15.0g
Distilled water	1.0Lt.
Final pH	7.0±0.1

Sterilize at 15 lbs pressure for 30 minutes

Procedure

1. Procure milk sample. Inoculate 1 ml of sample in 9 ml water blank to obtain 10^{-1} dilution. Transfer 1 ml of suspension to 9 ml water blank, it establishes a 10^{-2} dilution. From this second dilution, similarly transfer 1 ml quantities to water blank of 9 ml to provide dilution series ranging from 10^{-3} and 10^{-4} dilutions.
2. Transfer 1 ml suspension from each tube into petri dishes, and pour melted and cooled Tryptone glucose yeast extract agar.
3. Incubate the plates at 37°C for 24-48 hours.
4. Observe and count the number of colonies in each plate.
5. Meanwhile, inoculate 1 ml from highest dilution into MacConkey broth tube for the production of acid and gas.
6. Transfer MacConkey broth tube for production of acid and gas.
7. Place a single Durham tube in inverted position inside the broth containing tube and incubate for 24 hours.
8. Observe the acid and gas production.

Observations and Calculations

Calculate the results by multiplying average number of colonies per plate for a given dilution by the dilution factor to obtain viable count per ml of initial milk sample.

The acid and gas production constitutes a positive indication for the presence of coliforms.

S.No.	Dilution	Number of colonies
1	10^{-2}	
2	10^{-3}	
3	10^{-4}	

Total viable count per millilitre of milk:
Acid and gas produced in Mac Conkey broth tubes:

Precautions

i. The sample containers should be properly cleaned, dried and sterilized before use.

ii. Glassware, solutions, media etc used for the experiment should be properly sterilized.

iii. Petri plates should be kept upside down (inverted) in the incubator.

iv. Dilution bottles and petri plates should be labelled/marked properly.

Reference

Dubey R.C and Maheshwari D.K. (2004) Practical Microbiology. S. Chand & Company Ltd.

Questions

1. Why the temperature of agar should not exceed 45^{0}C at the time of pouring?
2. Why the results of SPC are expressed as cfu/ml or cfu/g

35

Determination of Microbial Quality of Milk by Methylene Blue Reduction Test

Introduction

Dye reduction tests show the comparative activity of the bacteria in milk, therefore are considered as rough indicator of the number of bacteria per millilitre of milk. These tests find wide application in dairy industry especially in grading of milk for manufacturing purposes. The small amount of equipment, materials and simplicity of method permit simultaneous testing of several samples. Since it combines speed and adaptability, this test may be used in milk plants, receiving stations, condensing plants, cheese factories and similar establishments to rapidly grade incoming milk.

Principle

It is based on the observation that the colour imparted to milk by a small amount of methylene blue disappears in a time, the length of which is dependant largely upon the number of microorganisms present in the milk. The change brought about is a chemical reduction of the dye due to action of bacteria, apparently by the utilization of oxygen dissolved in milk.

Requirements

1. Sterilized test tubes
2. Sterilized pipettes
3. Bunsen burner
4. Sterilized rubber stoppers
5. Water bath
6. Glass marker
7. Clock
8. Standard Methylene blue solution- Dissolve one tablet (8.8mg) of methylene blue thiocyanate in 200 ml of distilled water.

Procedure

1. Mix the milk thoroughly and transfer 10 ml of milk into a sterilized test tube with the help of sterilized pipette under aseptic conditions.
2. Add one ml of Methylene blue solution to the milk.
3. Close the test tube with sterilized stoppers.
4. Mix the contents thoroughly by inverting the tubes 3-4 times.
5. Place the test tube in water bath maintained at 37^{0}C.
6. Run the following controls with each batch of sample.

   Control-I : 10 ml of milk sample and 1 ml of tap water

   Control-II : 10 ml of milk sample and 1 ml of dye solution
7. Close the both tubes with rubber stopper.
8. Invert the tubes to mix the contents thoroughly.
9. Immerse the tubes in boiling water for 3 minutes.
10. Observe the test tubes in water bath after 30 minutes and then subsequently after every hourly interval.
11. Compare the tube with controls. Control-II reveals initiation of decolourization whereas Control-I indicates the completion of reaction.
12. Record the time at which complete decolourization is achieved and interpret the results as specified.

Observations

S.No.	Sample No.	Methylene blue reduction time (MBRT)

Interpretation

Methylene blue reduction time (hours)	Quality of Milk
5 and above	Very good
3 and 4	Good
1 and 2	Fair
½ and below	Poor

Precautions

i. Glassware should be properly cleaned and dried before use.

ii. Do not use chlorine or other chemical compounds as bactericidal agents on glassware.

iii. Do not touch the tip of pipette with milk while transferring dye.

iv. Use sterile test tubes for the test.

v. Always keep the dye solution in sterile light resistant glass containers in a cool dark place.

vi. Prepare fresh dye solution after a week.

Reference

Cappuccino J.C. and Sherman Natalie (IV-Edition) Microbiology A laboratory Manual. Addison Wesley Longman Inc, Replica Press Pvt. Ltd., Narela, Delhi.

Questions

1. What are the commonly used oxidation-reduction indicators for evaluating sanitary quality of milk.
2. What is the concentration of methylene blue thiocyanate in the recommended methodology?
3. Why control I & II are heated in boiling water?

36

Determination of Microbial Quality of Milk by Resazurin Test

Introduction

Resazurin has high tinctorial value (related to colour), imparting characteristic blue colour. Resazurin test gives us indirect information concerning bacterial content and /or activity. Resazurin is also sensitive to the reducing action of leucocytes. The results of this test are comparatively less affected by antibiotic residues in milk than methylene blue reduction test.

Principle

Reduction of dye is indicated by a gradual colour change through various shades of purple and mauve to full pink, at which point the resazurin completely changes to resorufin. This stage of reduction is irreversible. The second stage consisting of further reduction of resorufin to dihydroresorufin and characterized by a fading of the pink colour and is reversible.

Requirements

1. Sterilized test tubes
2. Sterilized rubber stoppers
3. Disc comparator with standard resazurin discs
4. Sterilized pipettes
5. Clock
6. Water bath
7. Sterile standard resazurin solution: Prepare 0.05 per cent (w/v) stock solution by dissolving resazurin in glass distilled water and store under refrigerator in amber coloured bottle. Prepare 0.005 per cent bench solution by diluting stock solution with distilled water and use with in 8 hours.

Procedure

1. Mix the milk thoroughly.
2. Take 10 ml of milk sample in a sterilized test tube using sterilized pipette under aseptic conditions.

3. Add 1 ml of bench solution of resazurin (0.005%) to the milk.
4. Close the test tube with the sterilized rubber stopper and mix the contents by inverting the tube 3-4 times.
5. Incubate the test tube in a water bath maintained at 37^{0}C.
6. Remove the test tube from water bath after 10 minutes.
7. Match the colour immediately with resazurin discs in comparator.
8. Place a control tube of milk without dye in comparator for comparison.
9. Interpret the results as specified.

Observations

S.No.	Sample No.	Disc Reading

Interpretation

Disc Reading	Quality of Milk
4 or higher	Satisfactory
3.5 to 1	Doubtful
0.5 to 0	Unsatisfactory

Precautions

i. Use sterilized glassware for the tests
ii. Transfer the sample and dye under aseptically near flame of Bunsen burner.
iii. Do not expose resazurin solution and tubes containing dye to direct sun light
iv. Protect sterilized equipments from dust and from contamination by unnecessary handling.

Reference

Kocher G.H. (2012) Practical teachings in Microbiology. Narendra Publishing House, New Delhi (India)

Questions

1. What are the other tests to determine the sanitary quality of milk?
2. What is the concentration of bench solution of resazurin in milk?

37

Preservation of Meat and Meat Products

Introduction

Due to inherent composition (high moisture and nutrients) of meat, it is highly perishable in nature. Hence, preservation of meat is essential to maintain its fresh quality. This is accomplished by various preservative techniques based on the principles of temperature control, moisture control and direct microbial inhibition.

Temperature Control

1. Chilling/Refrigeration

In this method, meat is stored at refrigeration temperature of 2 to 5°C. Chilling of animal carcasses is also carried out during ageing process, cutting transit, retail display and even in consumer household before ultimate use. Relative humidity during chilling is kept about 90%. The fresh meat has shelf life of 5-7 days at refrigerated temperature. The dressed poultry and fish are commonly chilled by immersing in cold water until the internal temperature of carcasses reach 4°C. However, large carcasses or cut up parts are chilled by forced air movement in chillers (<0°C)

2. Freezing

It is method of choice for the long-term preservation of meat. It ceases the microbial growth and retards the activity of enzymes. The muscles are frozen to a temperature of -10 to -30°C. The freezing point of muscles is approximately -0.5°C. The proper wrapping of fresh meat in polyethylene bags before freezing is a must to avoid freezer burn condition. The quality of frozen meat is most influenced by freezing rate. Shelf life of meat under freezing varies from 6 months to 1 year depending upon species, temperature, etc.

Methods of Freezing

i. **Plate freezing:** Meat packed in polythene bags are placed in trays, which remain in contact with metal freezer plates. Plate freezing is done at -10 to -30°C. This is commonly employed for small meat cuts and de-boned meat portions etc.

ii. **Blast freezing:** Meat cut up parts or small carcasses wrapped in polyethylene films are passed through tunnel of blast freezer at temperature -10 to -40°C and air velocity 30-170 meters / minute. Most practical and economical range is 760 meters/ minute at -30°C. The storage life of fresh buffalo meat, beef, mutton and chevon is approximately 6 months while that of pork and poultry is 4 months.

iii. **Still air freezing:** It is done in cabinet freezers at temperature -10 to -30°C. Properly packed meat is placed in cabinet freezer until it becomes rock hard.

iv. **Liquid Immersion:** The brine solution, glycerol or glycols are used as immersion medium. The meat pieces or cuts / fish / poultry, properly packed in LDPE bags are immersed until frozen.

v. **Cryogenic freezing:** Properly packed meat in polyethylene bags are sprayed with liquid nitrogen in the tunnel for 2.5 to 5 minutes. The meat is frozen and achieves temperature -10 to -40°C.

3. Thermal Processing

In this, two temperature regimes that of pasteurization and sterilization are generally used. Pasteurization is heating meat to a temperature range of 58-75°C for a specified time where most of the microorganisms are killed. Pasteurized meat is stored under refrigeration at 4-7 °C. Sterilization of meat is heating it at temperature above 100°C where all spoilage microorganisms are killed or their microbial cells are damaged beyond repair. Sterilized meat products have shelf life of 2 years in cans and one year in pouches at room temperature conditions.

4. Canning

It is another way of high temperature treatment for preservation accomplished by thermal sterilization of product held in hermetically sealed containers. Canning preserves substantially all sensory attributes viz. appearance, flavour and texture of meat products. The following series of steps are followed in conventional canning.

De-boned chunks (4 cm) are precooked (70 ºC for 15 minutes)

↓

Cans are filled first with gravy (condiments, tomatoes, spices, salt, etc.) then meat chunks and finally gravy

↓

Exhausting (Removal of air from cans for vacuum sealing)

↓

Steaming (done by double steamer machine)

↓

Retorting or thermal processing under pressure for sufficient duration to achieve commercial sterility

↓

Fast cooling to 30-40 ºC to give thermal shock to microorganisms

↓

Storage at room temperature (preferably 18±2 ºC)
Shelf life for two years in cans and one year in retort pouches

Moisture control

1. Drying or Dehydration

Drying can be accomplished under either sun or hot air oven. Meat chunks or small fishes are dried in oven or cabinet dryer at 70ºC for 12 to 16 hours.

2. Freeze-drying

It is removal of water from a food by sublimation from frozen state to vapour state under vacuum with low heat treatment.

Pre-freezing (Meat is frozen at -40 ºC) is done followed by low temperature in plate heat exchangers at 1 to 1.5 mm pressure of mercury (90-95% moisture loss) and Secondary drying at high temperature (4-8% bound water loss). Then Vacuum packaging is done. Freeze dried products can be rehydrated before use and it preserves most of nutritive and sensory acceptability.

3. Curing

It is the age-old method of preservation with the application of salt, nitrates, nitrites or other ingredients to meat. There are several methods of curing:

- **Dry curing:** Dry curing ingredients are rubbed on all over the surface of meat cut. e.g., curing of bacon

- **Pickle curing:** Meat cuts are dipped in curing ingredient solution (pickle) and it penetrate deep inside of the meat cuts. It is quicker method than dry curing, e.g. curing of pork shoulder.
- **Injection curing:** Concentrated solution of the curing ingredients are pumped into the meat through artery or injected by multiple needles in the muscular pork or beef e.g., curing of pork ham. It is also known as artery or stitch pumping.
- **Direct addition method:** Curing ingredients mix is directly added to fine comminuted meat, e.g., luncheon meat.

4. Smoking

Smoke constitutes a large number of wood degradation products such as aldehydes, ketones, organic acids, phenols, etc. which exert bacteriostatic and antioxidant effects besides imparting characteristic smoky flavour. Preservation of smoked meat is also due to surface dehydration and lowering of surface pH by smoke constituents. It is commonly carried out on cuts such as hams, butts and on many processed sausage items. Smoke is produced in specially designed 'smoke house' where combustion of sawdust or hard wood is done to a temperature of 300 °C at 35-40% relative humidity. The meat cuts, products or sausages are hanged in smoke house until the internal temperature of product reaches to 71°C (148°F).

Presently, liquid smoke is also used by the meat processors with several advantages of no installation cost, repeatability of process, time saving, minimum problem of carcinogenic compounds and little atmospheric pollution.

Preparation of liquid smoke

Pyrolysis of hard wood sawdust

↓

Smoke captured in water by drawing it counter current to water through an absorption tower

↓

Recycling of smoke water till desired concentration is achieved

↓

Ageing of smoke solution for polymerization and tar precipitation

↓

Filtration through cellulose pulp filter to remove hydrocarbons

↓

Dilution with acid and water

↓

Addition, immersion, or spray of smoke solution on the product for specified period

Formulation of Liquid Smoke solution

Liquid smoke	20 to 30 parts
Citric Acid or Vinegar	5 parts
Water	65-75 parts

Direct Microbial Inhibition

1. Irradiation

Electromagnetic radiations can ionize the molecule of their path. These radiations can destroy the micro-organisms by fragmenting their DNA molecules and causing ionization of inherent water. Gamma radiations are commonly used for food preservation. The dose of 5-10 kGy has bacteriostatic action and can enhance shelf life of fresh meat cuts and poultry products for 19 days. The higher dose of 40 kGy can sterilize pork, poultry and fish. Ultra violet rays of 2650 Å (Angstrom) are also used for surface sterilization of meat and meat products. There are three methods of radiation preservation.

Radappertization: It is equivalent to radiation sterilization of "Commercial sterility" as it is understood in the canning industry. The irradiation dose is 30-40 kGy.

Radicidation: It is equivalent to pasteurization of milk. The irradiation dose is 2.5 to 10 kGy.

Radurization: It is equivalent to pasteurization. It is a reduction of number of viable specific spoilage microbes by radiation. The typical dose of radurization is 0.75 to 2.5 kGy.

2. Chemical preservation

Following 'GRAS' chemicals are used in limited level as preservative because at higher level these chemicals directly affect consumer health. Different chemicals and their maximum permissible limit for using meat are: propionates / propionic acids (0.32%), sorbates / sorbic acid (0.2%), benzoates/ benzoic acid (0.1%), parabens (0.1%), sulphides/ SO_2 (200-300 ppm), ethylene/ propylene oxide (700 ppm), sodium diacetate (0.32%), dihydroacetic acid (65 ppm), sodium nitrite (150-200 ppm). Different organic acids like citric acid, lactic acid/ lactates and acetic acid at recommended level of 1-2% are also used for meat preservation. In addition, various **bacteriocins**, a group of natural food preservatives like; Niacin, Lacticin-481, Carnocin U-149, Pediocin- ACH, Sakacin A, Leucocin UAL, Helveticin J and Caseicin-80 etc. are obtained from different lactic acid bacteria and are recently in great demand for preservation of meat and meat products.

3. Hurdle Technology

Hurdle technology is a set of methods used for inactivation of microorganisms in food preservation. Hurdles are the food preservation factors which are combined to achieve certain stability or quality. The parameters like water activity, low temperature, acidification, redox-potential and different chemical preservatives are called hurdles. Hurdle Technology is used in development of shelf stable meat products, which requires no refrigeration for their storage. This is specifically important in minimally processed meat foods, which are developed on the principle of combining preservation methods following hurdle concept.

Reference

Frazier W.C. and Westhoff, D.C. (1988) Food Microbiology, Mc Graw Hill

Questions

1. What are the common methods of preservation of meat?
2. Enlist all the curing ingredients with their advantages used commonly for curing of meat?
3. What are the liquid smoke and its advantages?
4. What is the difference between radurization and radappertization?
5. Define a_w and its importance in preservation of food products?
6. What is commercial sterility?